FANTASTIKS
OF MATHEMATIKS

OF MATHEMATIKS

APPLICATIONS OF SECONDARY MATHEMATICS

CLIFF SLOYER

Janson Publications, Inc.

Providence, R.I.

Library of Congress Cataloging-in Publication Data

Sloyer, Clifford W.
    Fantastiks of mathematiks.

    Bibliography: p.
    Includes index.
        1. Mathematics——1961-   . 2. Mathematics——
Study and teaching (Secondary)   I. Title.
II. Title: Fantastics of mathematics.
QA37.2.S5735   1986   510   86-21326
ISBN 0-939765-00-4

Table of Contents

Subject Index   ix

Preface   xi

FANTASTIK 1:   **Clock the Block**   1
                     Mathematics used: *Combinations*

FANTASTIK 2:   **Top Copter**   9
                     Mathematics used: *Algebra, Inequalities*

FANTASTIK 3:   **Getting the Most from the Post**   11
                     Mathematics used: *Arithmetic-Geometric
                                          Mean Inequality (See
                                          Appendix I)*

FANTASTIK 4:   **Keep on Truckin'**   13
                     Mathematics used: *Algebra, Inequalities*

FANTASTIK 5:   **Is Inventory Mandatory?**   15
                     Mathematics used: *Arithmetic-Geometric
                                          Mean Inequality (See
                                          Appendix I)*

FANTASTIK 6:   **A Plane Gain**   19
                     Mathematics used: *Algebra, Systems of
                                          Equations*

FANTASTIK 7:   **A Slower Mower Maneuver**   21
                     Mathematics used: *Elementary Algebra*

FANTASTIK 8:   **A Chair Fare or A Table Fable**   24
                     Mathematics used: *Algebra, Graphing
                                          Inequalities*

FANTASTIK 9:   **Fertilizing Can Be Tantalizing**   27
                     Mathematics used: *Algebra, Graphing
                                          Inequalities*

FANTASTIK 10:   **Card Bard**   30
                     Mathematics used: *Elementary Algebra*

FANTASTIK 11:   **A Fishy Story**   32
                     Mathematics used: *Elementary Algebra and
                                          Probability*

FANTASTIK 12:   **A Replace Case**   36
                     Mathematics used: *Algebra, Inequalities*

FANTASTIK 13:   **The Mouse That Roared**   39
                     Mathematics used: *Algebra, Logarithms*

TABLE OF CONTENTS

FANTASTIK 14:   **The Child Prodigy**   41
Mathematics used: *Algebra, Elementary, Probability*

FANTASTIK 15:   **That #%&*?/ Mortgage Payment**   44
Mathematics used: *Algebra, Geometric Series*

FANTASTIK 16:   **A Smug Drug**   46
Mathematics used: *Algebra, Geometric Sequences and Series*

FANTASTIK 17:   **An Expected Die**   50
Mathematics used: *Elementary Probability*

FANTASTIK 18:   **Play a Sane Game**   53
Mathematics used: *Algebra, Elementary, Probability (Review Fantastik 17)*

FANTASTIK 19:   **Satellite Hite Site**   57
Mathematics used: *Trigonometry*

FANTASTIK 20:   **Pray Prey**   58
Mathematics used: *Algebra, Factoring Inequalities*

FANTASTIK 21:   **A Locatshun Problum**   61
Mathematics used: *Algebra, Absolute Value, Graphing*

FANTASTIK 22:   **A Dear Pier**   66
Mathematics used: *Arithmetic-Geometric Mean Inequality (See Appendix I)*

FANTASTIK 23:   **Judge Judges**   69
Mathematics used: *Algebra, Elementary Probability*

FANTASTIK 24:   **Slice the Price**   72
Mathematics used: *Quadratic Functions*

FANTASTIK 25:   **The Bore War**   76
Mathematics used: *Algebra, Elementary Probability (Review Fantastik 18)*

FANTASTIK 26:   **A New Pair of Genes**   78
Mathematics used: *Algebra, Elementary Probability*

FANTASTIK 27:   **Cues on Queues**   83
Mathematics used: *Algebra, Inequalities*

TABLE OF CONTENTS

FANTASTIK 28: **A Bloody Affair** 86
Mathematics used: *Algebra (Review Fantastiks 8 and 9)*

FANTASTIK 29: **Cite a Satellite** 88
Mathematics used: *Trigonometry*

FANTASTIK 30: **A Test Quest** 91
Mathematics used: *Elementary Probability, Binomial Tables*

FANTASTIK 31: **Tales of Scales** 92
Mathematics used: *Ratios*

FANTASTIK 32: **Play a Cool Pool** 99
Mathematics used: *Elementary Algebra, Similar Triangles*

FANTASTIK 33: **Relatively Relativity** 103
Mathematics used: *Pythagorean Theorem*

FANTASTIK 34: **Of Epidemic Proportions** 111
Mathematics used: *Algebra*

FANTASTIK 35: **Waste Makes Haste** 117
Mathematics used: *Calculus, Differential Equations*

FANTASTIK 36: **Bend the Trend or Raise the Razor** 120
Mathematics used: *Algebra, Elementary Probability, Matrix Algebra*

FANTASTIK 37: **A Chop Drop** 124
Mathematics used: *Distance Formula*

FANTASTIK 38: **A Smug Drug Revisited** 127
Mathematics used: *Difference Equations (See Appendix II, Review Fantastik 16)*

FANTASTIK 39: **A Lank Tank** 130
Mathematics used: *Difference Equations (See Appendix II)*

FANTASTIK 40: **Cobwebs** 134
Mathematics used: *Difference Equations (See Appendix II)*

APPENDIX I: **Arithmetic-Geometric Mean Inequality** 138

APPENDIX II: **Linear Difference Equations** 139

Some After Dinner Reading Suggestions 143

Subject Index

Absolute Value — FANTASTIK 21: *A Locatshun Problum*, p. 61.

Algebra — FANTASTIK 2: *Top Copter*, p. 9; FANTASTIK 4: *Keep on Truckin'*, p. 13; FANTASTIK 6: *A Plane Gain*, p. 19; FANTASTIK 8: *A Chair Fare or A Table Fable*, p. 24; FANTASTIK 9: *Fertilizing Can Be Tantalizing*, p. 27; FANTASTIK 12: *A Replace Case*, p. 36; FANTASTIK 13: *The Mouse That Roared*, p. 39; FANTASTIK 14: *The Child Prodigy*, p. 41; FANTASTIK 15: *That #%&*?/ Mortgage Payment*, p. 44; FANTASTIK 16: *A Smug Drug*, p. 46; FANTASTIK 18: *Play a Sane Game*, p. 53; FANTASTIK 20: *Pray Prey*, p. 58; FANTASTIK 21: *A Locatshun Problum*, p. 61; FANTASTIK 23: *Judge Judges*, p. 69; FANTASTIK 25: *The Bore War*, p. 76; FANTASTIK 26: *A New Pair of Genes*, p. 78; FANTASTIK 27: *Cues on Queues*, p. 83; FANTASTIK 28: *A Bloody Affair*, p. 86; FANTASTIK 34: *Of Epidemic Proportions*, p. 111; FANTASTIK 36: *Bend the Trend or Raise the Razor*, p. 120.

Algebra, Elementary — FANTASTIK 7: *A Slower Mower Maneuver*, p. 21; FANTASTIK 10: *Card Bard*, p. 30; FANTASTIK 11: *A Fishy Story*, p. 32; FANTASTIK 32: *Play a Cool Pool*, p. 99.

Arithmetic-Geometric Mean Inequality — FANTASTIK 3: *Getting the Most from the Post*, p. 11; FANTASTIK 5: *Is Inventory Mandatory?*, p. 15; FANTASTIK 22: *A Dear Pier*, p. 66; APPENDIX I: *Arithmetic-Geometric Mean Inequality*, p. 138.

Binomial Tables — FANTASTIK 30: *A Test Quest*, p. 91.

Calculus — FANTASTIK 35: *Waste Makes Haste*, p. 117.

Combinations — FANTASTIK 1: *Clock the Block*, p. 1.

Difference Equations — FANTASTIK 38: *A Smug Drug Revisited*, p. 127; FANTASTIK 39: *A Lank Tank*, p. 130; FANTASTIK 40: *Cobwebs*, p. 134; APPENDIX II: *Linear Difference Equations*, p. 139.

Differential Equations — FANTASTIK 35: *Waste Makes Haste*, p. 117.

Distance Formula — FANTASTIK 37: *A Chop Drop*, p. 124.

Geometric Sequences — FANTASTIK 16: *A Smug Drug*, p. 46.

Geometric Series — FANTASTIK 15: *That #%&*?/ Mortgage Payment*, p. 44; FANTASTIK 16: *A Smug Drug*, p. 46.

Graphing — FANTASTIK 21: *A Locatshun Problum*, p. 61.

Inequalities — FANTASTIK 2: *Top Copter*, p. 9; FANTASTIK 4: *Keep on Truckin'*, p. 13; FANTASTIK 12: *A Replace Case*, p. 36; FANTASTIK 27: *Cues on Queues*, p. 83.

Inequalities, Factoring — FANTASTIK 20: *Pray Prey*, p. 58.

SUBJECT INDEX

Inequalities Graphing — FANTASTIK 8: *A Chair Fare or A Table Fable*, p. 24; FANTASTIK 9: *Fertilizing Can Be Tantalizing*, p. 27.

Linear Difference Equations — APPENDIX II: *Linear Difference Equations*, p. 139.

Logarithms — FANTASTIK 13: *The Mouse That Roared*, p. 39.

Matrix Algebra — FANTASTIK 36: *Bend the Trend or Raise the Razor*, p. 120.

Probability — FANTASTIK 11: *A Fishy Story*, p. 32.

Probability, Elementary — FANTASTIK 14: *The Child Prodigy*, p. 41; FANTASTIK 17: *An Expected Die*, p. 50; FANTASTIK 18: *Play a Sane Game*, p. 53; FANTASTIK 23: *Judge Judges*, p. 69; FANTASTIK 25: *The Bore War*, p. 76; FANTASTIK 26: *A New Pair of Genes*, p. 78; FANTASTIK 30: *A Test Quest*, p. 91; FANTASTIK 36: *Bend the Trend or Raise the Razor*, p. 120.

Pythagorean Theorem — FANTASTIK 33: *Relatively Relativity*, p. 103

Quadratic Functions — FANTASTIK 24: *Slice the Price*, p. 72.

Ratios — FANTASTIK 31: *Tales of Scales*, p. 92.

Similar Triangles — FANTASTIK 32: *Play a Cool Pool*, p. 99.

Systems of Equations — FANTASTIK 6: *A Plane Gain*, p. 19.

Trigonometry — FANTASTIK 19: *Satellite Hite Site*, p. 57; FANTASTIK 29: *Cite a Satellite*, p. 88.

Preface

Framed within a facade of admitted "corn," these notes represent a serious attempt to make secondary teachers, prospective secondary teachers, and liberal arts students more aware of the interaction of mathematics and the real world. It is hoped that these notes will motivate secondary teachers, and through the teachers their students, as well as college students in non-technical areas, to a further study and enjoyment of mathematics. These notes have been used with several hundred students and the results have certainly been encouraging.

Very few practice problems appear in these notes. This was done purposely since a very important aspect in the use of these notes was to get students to work on their own models and applications of mathematics. For some, this meant going to the literature and looking for applications of elementary mathematics; others, however, actually built a model "from scratch." The results were either presented and discussed in class, or, if time did not permit, mimeographed for the students. The entire process was geared toward student involvement in the classroom.

In addition to examining a variety of applications, an attempt has been made to introduce the reader to some of the "newer" areas of mathematics, such as linear programming, dynamic programming, and geometric programming. Difference equations are introduced and frequently used. A computer was involved in many of the projects discussed. However, student knowledge of computer languages varied from "none" to FORTRAN to BASIC, etc., and hence few programs appear in these notes. In many cases, however, the benefit to be derived from the use of a computer should be obvious.

Many of the applications presented have already been used in secondary and college classrooms. The level of mathematics needed varies greatly. For example, CLOCK THE BLOCK has been used with secondary general mathematics students, while WASTE MAKES HASTE has been used in secondary and college calculus classes.

I would like to thank Norman Ricker for the many valuable suggestions he has made in revising the original set of notes and Dr. William Sacco of Tri-Analytics, Inc. for introducing me to dynamic programming. Thanks also go to Howard Hand for allowing me to use his notes on mathematics and music (see Fantastik 31).

It is hoped that these notes will have a positive effect in motivating students to a further study of mathematics.

Cliff Sloyer

Clock the Block

Mathematics Used: *Combinations*

Let us consider the following problem: We look at a "diagram" of four city blocks, where the lines represent streets (see Figure 1).

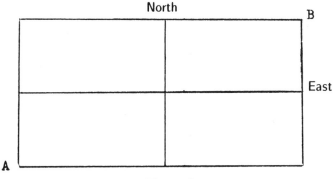

Figure 1

Suppose that the density of traffic on these streets and hence the times required to travel each block differ. The times in minutes are listed in Figure 2.

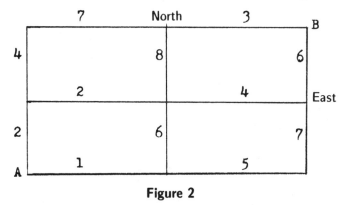

Figure 2

We wish to determine the fastest way of going from point A to point B subject to the following: At any corner we must go either east or north. Thus, we might take the path illustrated in Figure 3. It takes 17 minutes to travel the illustrated path. We shall designate this path by $ENEN$ using E for an east "step" and N for a north one.

Another possible path is shown in Figure 4. This path requires 16 minutes. We designate it as $NNEE$.

We observe that any path has 4 steps, 2 east steps and 2 north steps. We may represent the steps as follows:

——————— , ——————— , ——————— , ——————— ,

Step 1   Step 2   Step 3   Step 4

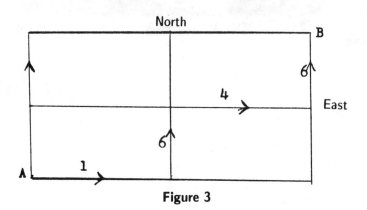

Figure 3

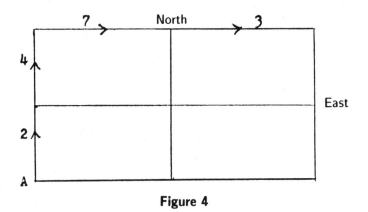

Figure 4

Once we have chosen the step numbers to be identified as east steps, the remaining ones must, of course, be north steps. Thus, any list of two E's and two N's determine a possible path. Hence, there are

$$\binom{4}{2} = \frac{4!}{2!2!} = 6$$

possible paths from A to B. We can list all of them as follows:

Path	Time in Minutes
(1) $EENN$	19
(2) $ENEN$	17
(3) $ENNE$	18
(4) $NNEE$	16
(5) $NENE$	15
(6) $NEEN$	14

We next examine the number of minutes required for each path and select that path, not necessarily unique, for which this number is smallest. In the above

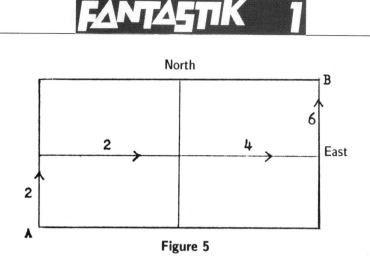

Figure 5

list, 14 minutes is the minimum, and hence we should choose the path $NEEN$ (see Figure 5).

Thus, we see that we can solve such problems and that the time spent is not unreasonable. As a matter of fact, in order to solve the above problem, we performed only 18 additions (3 per path) and 5 comparisons. One method for making such comparisons is the following: Compare the time for path 1 with the time for path 2; take the smaller of these two times and compare with the time for path 3; etc. Observe that the number of comparisons in this process is one less than the number of paths.

Now suppose that we have a diagram of 10^2 (10 by 10) city blocks. Then each path would consist of 10 north steps and 10 east steps. The number of possible paths is

$$\binom{20}{10} = \frac{20!}{10!10!} = 184,756.$$

In order to solve this larger problem in the same manner as the previous one, we would need to perform

19 (additions per path) × 184,756 (number of paths)

or 3,510,364 additions. We would also have to make 184,755 comparisons (recall that the number of comparisons required is one less than the number of paths). A dedicated student (with very little sleep) could solve this problem in a month. A computer, performing additions and comparisons at the rate of $10^5 = 100,000$ each second, would require approximately 37 seconds of computing time to solve the problem.

Let us now consider a diagram composed of 30^2 (30 by 30) city blocks. Each path would entail 30 east steps and 30 north steps. The number of possible paths would be

$$\frac{60!}{30!30!}.$$

This number exceeds 10^{17}. The number of additions per path is 59. Therefore, to determine the path of least time we would need to perform over 59×10^{17} additions and over 10^{17} comparisons. This total exceeds 10^{18}. A computer,

performing at the rate of 100,000 arithmetic operations per second (or about 3.1536×10^{12} operations per year), would take more than $10^5 = 100,000$ years to find the best path. The life expectancy of present computers is about 10 years.

Thus, we have the following table:

Size	No. of Operations	Human	Computer
2×2	23	< 10 min	$\sim$
10×10	3,695,119	$\approx$ mo	≈ 37 sec
30×30	$> 10^{18}$	$\sim$	$> 100,000$ yr

The computational procedure which we have been using is called *direct search*. The direct search procedure involves finding a value associated with **each** alternative action (in our case, the time associated with each path). There are many real-life problems in the physical, social, and management sciences that present more alternatives than our 30×30 block problem. We thus need a more "efficient" method for solving problems of this type. The method which we now illustrate is the basic idea underlying a relatively new (invented by Richard Bellman about 1950) branch of mathematics known as **dynamic programming**. The dynamic programming method will solve a 30×30 block problem using fewer than 3000 operations, an almost unbelievable reduction from the more than 10^{18} operations required by direct search.

Let us return to our 2×2 block problem illustrated in Figure 2. To simplify what follows, we introduce a coordinate system with the origin at A (see Figure 6).

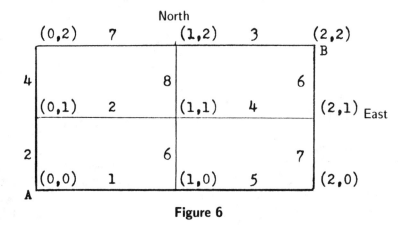

Figure 6

Let us start at B and work backwards as follows: If we arrive at corner $(1, 2)$ we have no choice but to go east, which will require 3 minutes. If we arrive at corner $(2, 1)$ we have no choice but to go north, which will require 6 minutes. Similarly, if we should arrive at corner $(2, 0)$, we must go north, which will require 13 minutes to reach B. If we should arrive at corner $(0, 2)$ we must go east, and in 10 minutes we reach B. We indicate these numbers and directions in Figure 7.

If we arrive at corner $(1, 1)$, we do have a choice of E or N. Using E would require 10 minutes; using N would require 11 minutes. Thus, if we arrive at

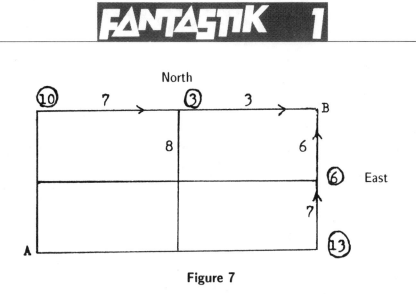

North

Figure 7

corner $(1, 1)$, then the best we can do is to use E, and it is going to take 10 minutes to reach B. We indicate this in Figure 8.

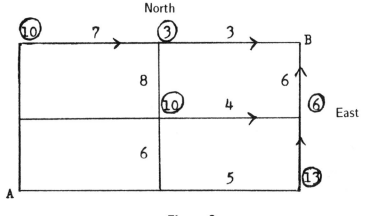

North

Figure 8

If we should arrive at corner $(1, 0)$, we have a choice of E or N. A choice of N would yield a trip of at least 16 minutes to B, while a choice of E would result in a trip of 18 minutes. Thus, if we should arrive at corner $(1, 0)$, the best we can do is to use N and it is going to take 16 minutes to reach B. We indicate this as shown in Figure 9.

We continue to proceed in this manner, placing a number at each corner and indicating the direction we should choose with an arrow. We thus obtain the diagram in Figure 10.

The solution to our problem is now obvious. We start at A and follow the arrows, which, as we can see, yields the path $NEEN$. The number inserted at corner A tells us immediately the minimum number of minutes required in order to go from A to B. Notice that our solution, $NEEN$, is the same as the one computed previously using direct enumeration. Moreover, the new method is much more efficient. At every corner, we were required to do no more than two

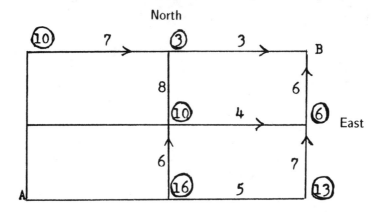

Figure 9

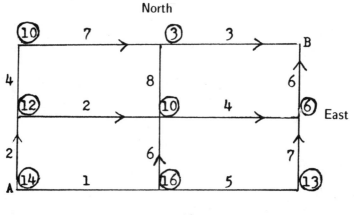

Figure 10

additions and one comparison (3 operations). Thus, for example, the 30×30 diagram, which has $31 \times 31 = 961$ corners, would require fewer than 3000 operations in order to find the best path. Time (on a computer) has been reduced to $3/100$ sec.

We thus have the following tables:

Size	No. of Operations	Human	Computer
	Direct Search		
2×2	23	< 10 min	$\sim$
10×10	3,695,119	$\sim$ mo	~ 37 sec
30×30	$> 10^{18}$	$\sim$	> 100,000 yrs
	Dynamic Programming		
2×2	< 27	< 10 min	$\sim$
10×10	< 363	~ 10 min	$< \frac{4}{1000}$ sec
30×30	< 3000	~ 75 min	$< \frac{3}{100}$ sec

Practice Problem 1: Use the above technique to find the minimum number of minutes required to go from A to B (see figure below). Recall that at each corner, one must go east or north.

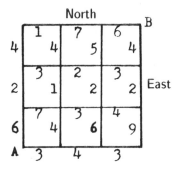

Practice Problem 2: Use the above technique to find the minimum number of minutes required to go from A to B (see figure below). Again recall that at each corner, one must go east or north.

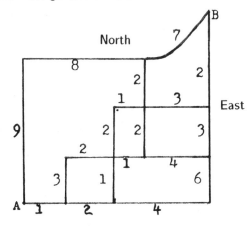

Practice Problem 3: A water line is to be constructed in the basement of a building. The line can follow numerous paths from the intake point A to point

B, and these are indicated in the figure below. Because the lengths of copper tubing and the obstructions differ, different paths result in different construction costs. The cost of required copper tubing appears beside each section of the path. An obstruction is denoted by the symbol $\otimes$ and the cost of removing this obstruction is given below the symbol. Which path should be chosen to minimize the cost?

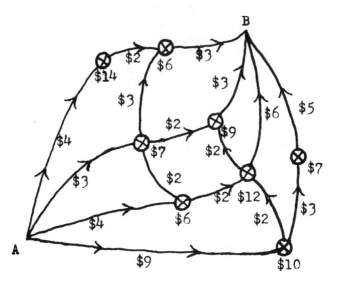

Practice Problem 4:   Give an example which shows the following: If one removes the restriction that at each corner one must go east or north, then the particular technique of this section need not yield an "optimal" path.

8

Top Copter

Mathematics Used: *Algebra, Inequalities*

A certain Sikorsky helicopter can carry cargo internally within its fuselage or externally slung beneath its fuselage. When carried externally the cargo induces a drag on the helicopter which yields less average speed than when carrying cargo internally. However, external loading and unloading usually require less time than internal loading and unloading.

We consider here a cargo which can be loaded by either of the methods described above, but not by a combination (e.g., half internally and half externally). Suppose the cargo is to be carried for a distance of 80 miles and delivery time is the prime factor. Assume the following data:

	Average Speed (mph)	Loading Time (h)	Unloading Time (h)
Internal Load	140	$\frac{1}{4}$	$\frac{1}{4}$
External Load	100	$\frac{1}{12}$	$\frac{1}{12}$

Table 1

Observe that the flight time for an internal load is 80/140 hours, while the flight time for an external load is 80/100 hours. Thus, the total delivery time for an internal load is

$$\underbrace{\frac{1}{4}}_{\text{load}} + \underbrace{\frac{80}{140}}_{\text{flight}} + \underbrace{\frac{1}{4}}_{\text{unload}} = \frac{15}{14} \quad (\approx 1.07 \text{ hours}).$$

The total delivery time for an external load is

$$\underbrace{\frac{1}{12}}_{\text{load}} + \underbrace{\frac{80}{100}}_{\text{flight}} + \underbrace{\frac{1}{12}}_{\text{unload}} = \frac{29}{39} \quad (\approx .97 \text{ hours}).$$

Hence, to ship this cargo for 80 miles, the external load is preferable.

Now, suppose we wish to ship the same cargo for 200 miles. In this case, the flight time for an internal load is 200/140, while the flight time for an external load is 200/100. Thus, the total delivery time for an internal load is

$$\underbrace{\frac{1}{4}}_{\text{load}} + \underbrace{\frac{200}{140}}_{\text{flight}} + \underbrace{\frac{1}{4}}_{\text{unload}} = \frac{27}{14} \quad (\approx 1.9 \text{ hours}).$$

The total delivery time for an external load is

$$\underbrace{\frac{1}{12}}_{\text{load}} + \underbrace{\frac{200}{100}}_{\text{flight}} + \underbrace{\frac{1}{12}}_{\text{unload}} = \frac{13}{6} \quad (\approx 2.2 \text{ hours}).$$

Hence, in this case, the internal load is preferable.

One can gain more information about "what's goin' on" by generalizing this process in the following way: Suppose that a helicopter is to carry this cargo for a distance of D miles. We continue to use the data in Table 1.

Observe that the flight time for an internal load is $D/140$, while the flight time for an external load is $D/100$. Thus, the total delivery time for an internal load is

$$\underbrace{\frac{1}{4}}_{\text{load}} + \underbrace{\frac{D}{140}}_{\text{flight}} + \underbrace{\frac{1}{4}}_{\text{unload}},$$

while the total delivery time for an external load is

$$\underbrace{\frac{1}{12}}_{\text{load}} + \underbrace{\frac{D}{100}}_{\text{flight}} + \underbrace{\frac{1}{12}}_{\text{unload}}.$$

The external load is preferred over the internal load whenever

$$\frac{1}{12} + \frac{D}{100} + \frac{1}{12} < \frac{1}{4} + \frac{D}{140} + \frac{1}{4}$$

or

$$\frac{D}{100} + \frac{1}{6} < \frac{D}{140} + \frac{1}{2}$$

or (multiplying both sides by 2100)

$$21D + 350 < 15D + 1050$$

or

$$6D < 700$$

or

$$D < \frac{700}{6} \quad (\approx 117 \text{ miles}).$$

That is, one would load externally if $D < 117$ miles and internally if $D > 117$ miles. (If $D = 700/6$, each method requires the same time.)

Practice Problem: For a distance D, which loading method is preferred, given the following data?

	Average Speed (mph)	Loading Time (h)	Unloading Time (h)
Internal Load	200	$\frac{1}{4}$	$\frac{1}{4}$
External Load	150	$\frac{1}{8}$	$\frac{1}{8}$

FANTASTIK 3

Getting the Most From the Post

Mathematics Used: *Arithmetic-Geometric Mean
Inequality (See Appendix I)*

The post office has limited the size of rectangular boxes sent through the mail by stating that "length plus girth cannot exceed 100 inches." (Recall that *girth* is the perimeter of a cross-section.) Our problem is to determine the dimensions of the largest (in terms of volume) rectangular box which can be sent through the U.S. mails.

Let l, w, and h denote the length, width, and height, respectively, of a rectangular box.

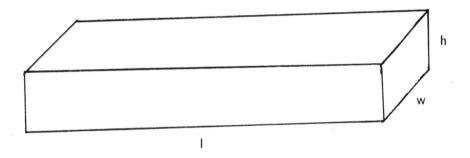

The girth is $2w + 2h$ so that the length plus girth is given by

$$l + 2w + 2h.$$

The value V is given by
$$V = lwh.$$

Our problem is to find positive numbers l, w, and h, so that

$$l + 2w + 2h \leq 100$$

and $V = lwh$ is as large as possible.

It should be clear that if V is to be as large as possible, then we must have

$$l + 2w + 2h = 100,$$

for suppose $l + 2w + 2h = 90$. Then by increasing l by 10 inches, we have

$$l + 2w + 2h = 100$$

and have *increased* the volume.

Thus, we want to find positive numbers l, w, and h, so that

$$l + 2w + 2h = 100 \tag{1}$$

11

and $V = lwh$ is as large as possible. Using the idea that

$$\frac{a_1 + a_2 + a_3}{3} \geq \sqrt[3]{a_1 a_2 a_3}, \quad ^*$$

we have

$$\frac{l + 2w + 2h}{3} \geq \sqrt[3]{4lwh} \qquad (2)$$

or

$$\frac{100}{3} \geq \sqrt[3]{4V} \qquad (3)$$

or

$$\frac{1}{4}\left(\frac{100}{3}\right)^3 \geq V. \qquad (4)$$

Thus, it is impossible to construct a rectangular box within the specified limits which has a volume *greater* than $\frac{1}{4}(\frac{100}{3})^3$. Moreover, in order to obtain this *optimal* value, we must have equality in (4), and hence equality in (2), which implies

$$l = 2w = 2h.^*$$

Substituting into (1), we get

$$3l = 100$$

or

$$l = \frac{100}{3},$$

so that the dimensions of our "largest rectangular box" are given by

$$\text{length} = \frac{100}{3} \text{ inches}$$
$$\text{width} = \frac{50}{3} \text{ inches}$$
$$\text{height} = \frac{50}{3} \text{ inches.}$$

Practice Problem: In certain areas, the U.S. mails limited the size of a rectangular box by stating that "length plus girth cannot exceed 72 inches." Under this requirement, what are the dimensions of the largest box which can be sent through the mails?

---

*See Appendix I.

Keep on Truckin'

Mathematics Used: *Algebra, Inequalities*

On November 24, 1973, a Los Angeles Times Service bulletin stated that Alan Glassenapp, staff engineer in the GMC Division of Research and Development, insisted that "big trucks and buses will get an average of 5 per cent fewer miles per gallon traveling at 55 MPH than at 50." He also stated that "in terms of cost effectiveness, the higher speed makes sense."

If one accepts the above quote and the 5 per cent figure, then it is clear that energy will be saved by traveling at 50 MPH rather than at 55 MPH. We now wish to analyze what will happen to the costs involved for a trip of distance D miles. Using the following notation,

$x =$ miles per gallon at 50 MPH
$f =$ cost (in dollars) per gallon of fuel
$c =$ cost (in dollars) per mile of all operating costs
     exclusive of fuel and wages for driver
$w =$ hourly wage (in dollars) of driver

we have

$$T_{50} = \frac{D}{x} \cdot f + \frac{D}{50} \cdot w + Dc$$

and

$$T_{55} = \frac{D}{.95x} \cdot f + \frac{D}{55} \cdot w + Dc,$$

where T_i denotes the total cost when traveling at i miles per hour.

In order to compare these two costs we consider

$$T_{55} - T_{50} = \frac{D}{x}f\left(\frac{1}{.95} - 1\right) + Dw\left(\frac{1}{55} - \frac{1}{50}\right)$$

or

$$T_{55} - T_{50} = \frac{D}{x}f\left(\frac{1}{19}\right) - Dw\left(\frac{1}{550}\right).$$

It follows that

$$T_{55} > T_{50}$$

whenever

$$\frac{D}{x}f\left(\frac{1}{19}\right) > Dw\left(\frac{1}{550}\right)$$

or

$$\frac{f}{wx} > \frac{19}{550}. \tag{1}$$

From (1) we can make observations such as the following:

    (I) A comparison of the total costs does not involve the distance D or the cost c.

(II) Suppose $f = .54$ and $x = 6$ (these figures are based on information received in 1974 from a local truck driver). Then (1) becomes

$$\frac{.54}{6w} > \frac{19}{550}$$

or (to two decimal places)

$$2.61 > w.$$

That is, in this case, the total cost at 50 MPH is less than at 55 MPH whenever the driver is earning less than $2.61 per hour.

(III) Suppose $x = 6$ and the driver's wages are $4.50 per hour. Then (1) becomes

$$\frac{f}{(4.5)(6)} > \frac{19}{550}$$

or (to two decimal places)

$$f > 0.93.$$

That is, in this case, the total cost at 50 MPH is less than at 55 MPH whenever fuel costs more than $.93 per gallon.

There is an assumption being made here that "c" is the same at 50 MPH as at 55 MPH. Although no research was discovered on this item, local truck drivers indicated that they thought there would be little if any difference.

Let us now make a similar comparison of costs based on information received in 1981 from a local truck driver.

(i) If $f = \$1.36$ and $x = 6$, then (1) becomes

$$\frac{1.36}{6w} > \frac{19}{550}$$

or (to two decimal places)

$$6.56 > w.$$

That is, in this case, the total cost at 50 MPH is less than at 55 MPH whenever the driver is earning less than $6.56 per hour.

(ii) Suppose $x = 6$ and the owner of a truck company must pay a driver $12.50 per hour. Then (1) becomes

$$\frac{f}{(12.5)(6)} > \frac{19}{550}$$

or (to two decimal places)

$$f > 2.59.$$

Practice Problem: Suppose that big trucks and buses get an average of 10 per cent fewer miles per gallon traveling at 60 MPH than at 55 MPH. Set up a model and analyze cost effectiveness based on the 1974 information in (II) and also on the 1981 information given in (i).

14

Is Inventory Mandatory?

Mathematics Used: *Arithmetic-Geometric Mean Inequality*
(See Appendix I).

One frequently hears the expression "low costs through high output." A person who compares the price of a new house in a development to the price of a new custom-built house is aware of this idea. However, there are many business activities in which some costs increase as the manufactured output increases. For example, there may be a need for financing the business activity, in which case, there is an interest cost. Also, if large quantities are manufactured and not sold immediately, there may be an inventory cost involving the cost of storage facilities, insurance, etc. For example, suppose that a company in the steel fabricating business has to supply 1600 units of a certain item to a customer during the coming year. This item is unique in the sense that it is purchased only by this customer. Now, there are certain costs involved which do not depend on the number of items to be made: obtaining proper tools from storage, setting up the tools, test production of a few sample items, secretarial time, etc. Let us assume that these **setup costs** amount to $100. The cost of producing **one unit** (labor, material, electricity, repair of machines, etc.) is $3. In addition, there are **inventory costs** (cost of storage facilities, insurance, etc.) which are determined as follows:

Total Inventory Costs = $2 × (average number of units in stock during year).

First, let us suppose that the manager decides to produce all 1600 units immediately. We assume that these items are sold at a uniform rate during the year. Since the stock will be depleted at the end of the year, the average number of items held in stock during the year is 800. The total cost of this business activity is thus given by

$$C = \overbrace{100}^{\substack{\text{Setup}\\\text{Cost}}} + \overbrace{2(800)}^{\substack{\text{Inventory}\\\text{Cost}}} + \overbrace{3(1600)}^{\substack{\text{Production}\\\text{Cost}}}$$

or

$$C = \$6500.$$

Now suppose that the manager decides to produce 800 immediately and another batch of 800 six months from now. S/he now has to pay for *two* setups. In addition, there is an average of 400 items in stock during the year. The total cost C of this business activity is now given by

$$C = 2(100) + 3(1600) + 2(400)$$

or

$$C = \$5800.$$

Observe that the total cost of the required production has decreased by running batches, rather than one. We might now consider (and possibly compute) what happens to the total cost if 4 batches of 400 are run, 8 batches of 200 are run, 5 batches of 320 are run, etc. The diagrams in Figure 1 are a conceptual aid in considering the many production schedules that are available.

In particular, we are seeking to discover a process which will minimize the total cost and yet still produce the required number of items. The richness

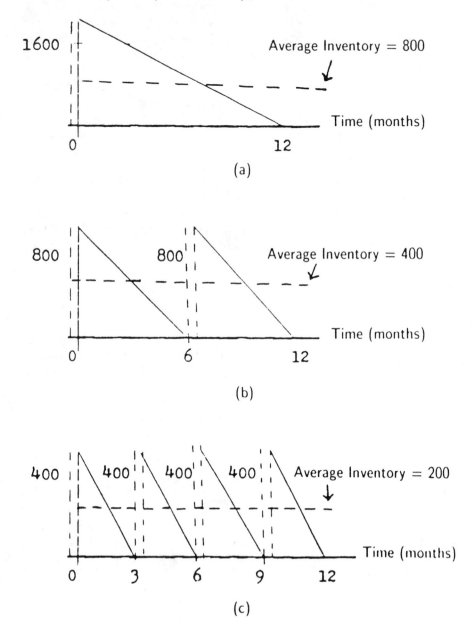

Figure 1

of mathematics, however, enables us to determine an optimal policy without making such numerous considerations. We shall assume that the same number of items is produced at each run. Moreover, we shall assume that the demand for these items remains constant, so that the time between production runs is also constant.

Let x denote the number of items in the batch produced during each run. If N runs are made, then we want $Nx = 1600$ and hence the number of runs is given by $1600/x$. Thus, we have

Number of production runs in year $= \dfrac{1600}{x}$

Total setup costs $= 100\left(\dfrac{1600}{x}\right)$

Average number of items in stock $= \dfrac{x}{2}$

Total inventory costs $= 2\left(\dfrac{x}{2}\right)$.

The total cost involved, denoted C, is thus given by

$$C = 100\left(\frac{1600}{x}\right) + 2\left(\frac{x}{2}\right) + 3(1600)$$
$$C = \frac{160,000}{x} + x + 4800.$$

The \$4800 is a fixed cost over which one has no control. We thus look at

$$\frac{160,000}{x} + x.$$

Using the basic arithmetic-geometric mean inequality,

$$\frac{\dfrac{160,000}{x} + x}{2} \geq \sqrt{160,000}$$
$$\frac{160,000}{x} + x \geq 800.$$

That is, the costs that we can control cannot be less than \$800. Moreover, in order to attain this minimal amount one must have

$$\frac{160,000}{x} = x$$
$$160,000 = x^2$$
$$x = 400.$$

Thus, the **optimal procedure** is to produce 400 items in each run, which means a total of 4 runs a year, one every 3 months.

The ideas used here form the basis of a relatively new area of mathematics known as **geometric programming**.

Practice Problem 1: A fabric manufacturer must supply a customer with 12,000 yards of a certain cloth during the coming year. There is a constant demand for the cloth during the year and the manager wants to produce the same number of yards at each production run. We have the following information:

   (i) Setup costs—$300
  (ii) Inventory costs—$.8$x$ (average number of yards in stock during the year)
 (iii) Production costs—$3 per yard.

How many yards should be made during each production run in order to minimize the total cost of this business activity?

Practice Problem 2: Generalize the example given, using the following notation.

$D =$ annual demand
$S =$ setup cost
$I =$ number which multiplied by average number of items in
      stock during year yields annual inventory costs
$C =$ cost of producing one item
$p =$ number of items produced in each run
$n =$ number of runs per year.

Show that an "optimal" policy is to take

$$p = \sqrt{\frac{2DS}{I}} \qquad \left(\text{or } n = \sqrt{\frac{DI}{2S}} \right).$$

A Plane Gain

Mathematics Used: *Algebra, Systems of Equations*

One day the president of On Time Airlines reads in the *Wall Street Journal* that Bald Mountain Airlines has ordered 13 new planes from Commuter Aircraft, Inc. S/he knows that three types of planes are produced by Commuter and that plane A sells for 1.1 million, plane B for 1.3 million, and plane C for 1.8 million. The president of On Time wants to know how many planes of each type were ordered. (Why might s/he want to know this?)

We first summarize the above data in the following table:

Plane	Cost(in millions)
A	1.1
B	1.3
C	1.8

Letting x denote the number of plane A's ordered, y the number of plane B's and z the number of plane C's, we have

$$\left.\begin{aligned}1.1x + 1.3y + 1.8z &= 16.5\\ x + y + z &= 13\end{aligned}\right\} \tag{1}$$

However, there is also a **condition** attached, namely, $x, y,$ and z **must be non-negative integers**.

Now, equation (1) may be rewritten as

$$\left.\begin{aligned}11x + 13y + 18z &= 165\\ x + y + z &= 13.\end{aligned}\right\} \tag{2}$$

Multiplying the second equation in (2) by -11 and adding the two equations, we obtain

$$2y + 7z = 22. \tag{3}$$

Dividing both sides of (3) by 2, we have

$$y + 3z + \frac{1}{2}z = 11$$

or

$$\frac{z}{2} = 11 - y - 3z.$$

Now, since y and z are to be integers, it must be the case that $11 - y - 3z$ is an integer. Thus, we write

$$\frac{z}{2} = n, \quad n \text{ an integer}$$

19

or

$$z = 2n. \tag{4}$$

Substitution of (4) into (3) yields

$$2y + 14n = 22$$

or

$$y = 11 - 7n. \tag{5}$$

Substitution of (4) and (5) into the second equation of (2) yields

$$x + (11 - 7n) + (2n) = 13$$

or

$$x = 2 + 5n.$$

Summarizing, we have

$$x = 2 + 5n$$
$$y = 11 - 7n$$
$$z = 2n$$

where n is an integer. We now construct the following table:

n	x	y	z
0	2	11	0
1	7	4	2

In the construction of the above, note that we could not use an integer n which is 2 or greater since, in that case, y would be negative. Also, we could not use a negative integer n since, in that case, z would be negative.

Thus, there are two possible solutions. Bald Mountain Airlines has ordered either

(i) 2 type A planes, 11 type B planes, no type C planes

or

(ii) 7 type A planes, 4 type B planes, and 2 type C planes.

Practice Problem: With the costs per plane given in this section, how many of each type were purchased by Bald Mountain Airlines if the total cost is 19.3 million?

A Slower Mower Maneuver

Mathematics Used: *Elementary Algebra*

A certain lawn measures approximately 30′ by 40′. In the past, two different mowing techniques have been used. First, one mowed lengthwise, made a complete turn, again mowed lengthwise, etc., as in the following figure.

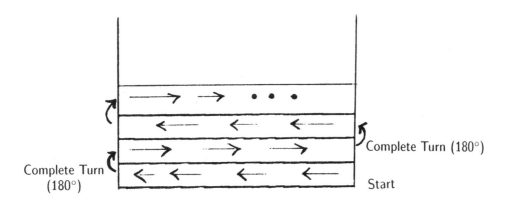

Second, one went around the perimeter, around the resulting perimeter, etc., as illustrated in the following figure:

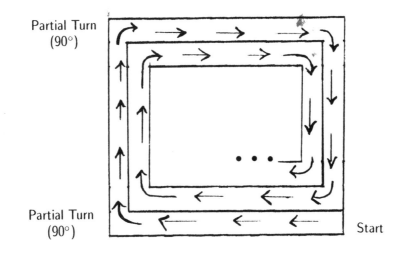

The question is, "Which method is faster if we assume that the person mowing walks at the same speed?"

Now, in either case, one is going to mow exactly the same amount of grass; and, hence, the problem reduces to a consideration of the time that is involved in making the various turns. Let T_c denote the time required for making a complete

turn (180°) and T_p denote the time required for making a partial turn (90°). We will assume that our mower is capable of mowing a strip 2' wide.

With the first method, it should be clear that 14 complete turns are made, 7 on the left side and 7 on the right side, as illustrated in the figure below.

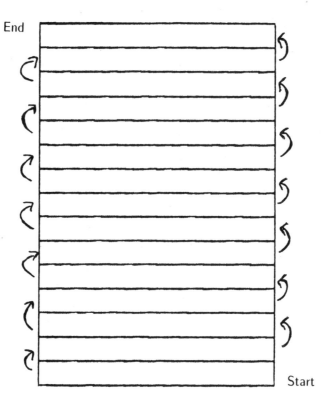

Thus, using this method, the time for turns is given by $14T_c$.

In order to determine the time for turns using the second method, let us consider one perimeter as in the figure below:

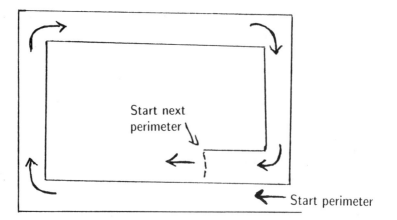

We see that each perimeter requires 4 partial turns before we begin the next perimeter. Suppose we have already mowed over 6 perimeters and hence, made 24 partial turns and reduced the size of the original lawn by 12′ on each side. What remains to mow is given in the following figure.

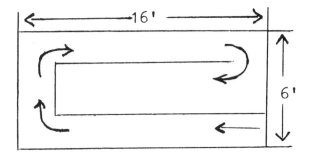

Mowing this path will require two additional partial turns and one complete turn as shown. The total time for turns using the second method is thus

$$26T_p + T_c.$$

It would be an "optimal" policy to use the second method whenever

$$26T_p + T_c < 14T_c$$
$$26T_p < 13T_c$$
$$2T_p < T_c.$$

That is, the second method should be used if the time for a complete turn is more than twice the time required for a partial turn.

(In checking with a lawn mower, the author found that the time for a complete turn was three times as much as the time required for a partial turn; i.e., $T_c = 3T_p$. Thus, for me, the better of the two given methods is the second method.)

Practice Problem 1: What would the results of this section be if the mower width were 20 inches?

Practice Problem 2: What would the results of this section be if the grass area measured 60′ by 100′ and the mower width was 2′?

Practice Problem 3: Generalize the ideas of this section by considering a grass area of $W' \times L'$ where W' is the width and L' is the length. Assume the mower width is M'. (You may assume that M divides W.)

A Chair Fare or A Table Fable

Mathematics Used: *Algebra, Graphing Inequalities*

Consider the following problem. A small manufacturing company produces 2 items, say chairs and tables. Each chair requires 3 feet of lumber, and each table requires 7 feet of lumber. A chair requires 2 labor-hours of time, and a table requires 8 labor-hours of time. The profit on a chair is $1, and the profit on a table is $3. If 420 feet of lumber and 400 labor-hours of time are available, how many of each item should the company produce in order to make its profit as large as possible?

Let x denote the number of chairs to be produced and y the number of tables. The information concerning required lumber and available lumber yields the inequality

$$3x + 7y \leq 420.$$

The information concerning required time and available time yields the inequality

$$2x + 8y \leq 400.$$

If P denotes the profit on x chairs and y tables, then one has

$$P = x + 3y.$$

We can thus reformulate our problem as follows: Find non-negative numbers x and y such that

$$3x + 7y \leq 420,$$
$$2x + 8y \leq 400,$$

and $P = 2x + 3y$ is as large as possible.

It is easy to show that the graph of the inequality

$$3x + 7y \leq 420$$

is given by the shaded area in Figure 1 and the graph of the inequality

$$2x + 8y \leq 400$$

is given by the shaded area in Figure 2. Hence, the set of points (x, y), with x and y non-negative, which satisfy both inequalities is given by the shaded area in Figure 3. The set of points (x, y) in the shaded region illustrated in the figure is referred to as the **feasible region**. That is, if (x, y) is a point which is not in the shaded region, then production of x chairs and y tables would be impossible; otherwise, such production is possible.

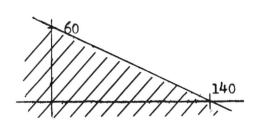

Figure 1

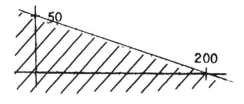

Figure 2

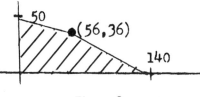

Figure 3

Now let us return to the profit function

$$P = x + 3y.$$

Observe that for any value of P, this equation represents a straight line which crosses the y-axis at $P/3$ and which has a slope of $-1/3$. Figure 4 illustrates the graph of this straight line for various values of P. The direction of increasing profit P is illustrated in Figure 5.

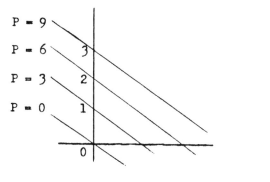

Figure 4

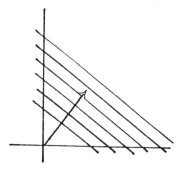

Figure 5

Let us now indicate the graph of a profit line, together with the graph in Figure 3, which indicates the feasible solutions (Figure 6). Observing the direction of increasing profit P, it should be clear that if we want (x, y) to be a feasible solution and P to be as large as possible, then we want the profit line illustrated in Figure 7. Thus the manufacturer should produce 56 chairs and 36 tables to maximize the profit P. Moreover, the profit P is easily calculated to be $164.

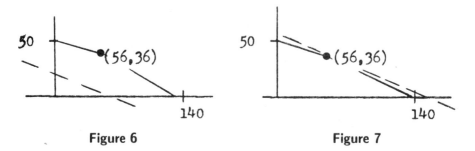

Figure 6   Figure 7

Problems of the type presented in this section are known as Linear Programming problems.

Practice Problem: A small manufacturing company produces 2 items, say chairs and tables. Each chair requires 3 feet of lumber, and each table requires 7 feet of lumber. A chair requires 2 labor-hours of time, and a table requires 8 labor-hours of time. The profit on a chair is $2, that on a table is $3. If 126 feet of lumber and 120 labor-hours of time are available, how many of each item should the company produce to make the profit as large as possible?

Fertilizing Can Be Tantalizing

Mathematics Used: *Algebra, Graphing Inequalities*

Let us consider the following problem: A certain company manufactures two kinds of lawn fertilizer, regular and deluxe. A bag of their regular fertilizer contains 3 lbs of nitrogen, 4 lbs of phosphoric acid, and 1 lb of potash. A bag of their deluxe fertilizer contains 2 lbs. of nitrogen, 6 lbs of phosphoric acid, and 3 lbs of potash. It has been determined that a certain lawn requires at least 10 lbs of nitrogen, 20 lbs of phosphoric acid, and 7 lbs of potash. The cost of a bag of regular fertilizer is $3, and the cost of a bag of deluxe fertilizer is $4. The problem, of course, is to determine how many bags of each should be purchased in order to provide effective fertilization and minimize the cost.

Let x denote the number of bags of regular fertilizer and y the number of bags of deluxe fertilizer to be purchased. Let C denote the cost of such a purchase. Our problem becomes the following: Find non-negative numbers x and y such that

$$\left. \begin{array}{r} 3x + 2y \geq 10, \\ 4x + 6y \geq 20, \\ x + 3y \geq 7, \end{array} \right\} \qquad (1)$$

and such that

$$C = 3x + 4y$$

is as small as possible.

It is readily established that the graph of the points (x, y) with x and y non-negative which satisfy all three inequalities in (1) is shown in Figure 1, which, therefore, is a graph of the feasible region for our problem.

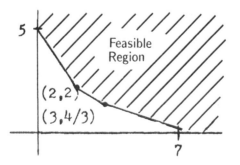

Figure 1

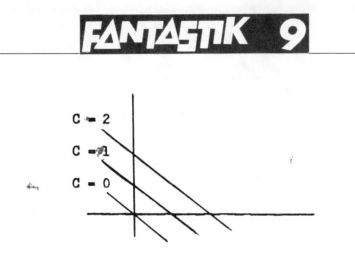

Figure 2

We now return to the cost function

$$C = 3x + 4y.$$

Observe that for any value of C, this equation represents a straight line which crosses the y-axis at $C/4$ and has a slope of $-3/4$. Figure 2 illustrates the graph of this straight line for various values of C. The direction of decreasing cost C is illustrated in Figure 3.

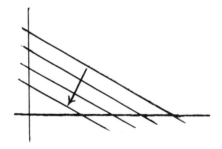

Figure 3

Let us now indicate the graph of a cost line, together with the graph in Figure 1, which indicates the feasible solutions (see Figure 4). Observing the direction of decreasing cost C, it is clear that if we want (x, y) to be a feasible solution and C to be as small as possible, we should choose the cost line illustrated in Figure 5. Thus, if one purchases 2 bags of regular fertilizer and 2 bags of deluxe fertilizer, one will have the necessary ingredients; in addition, the cost, namely $14, is as small as possible.

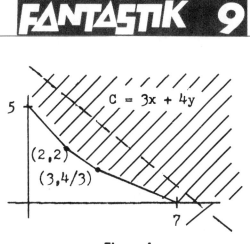

Figure 4

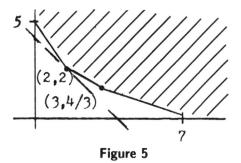

Figure 5

Problems of the type considered in this section (as well as in Fantastik 8) are known as **Linear Programming** problems.

Practice Problem: A certain company manufactures two kinds of lawn fertilizer, mix A and mix B. A bag of mix A contains 4 lbs of nitrogen, 2 lbs of phosphoric acid, and 1 lb of potash. A bag of mix B contains 3 lbs of nitrogen, 2 lbs of phosphoric acid, and 4 lbs of potash. The cost of a bag of mix A is $8; the cost of a bag of mix B is $6. A certain lawn requires at least 18 lbs of nitrogen, 10 lbs of phosphoric acid, and 8 lbs of potash. Determine how many bags of each should be purchased in order to provide effective fertilization and minimize the cost.

Card Bard

Mathematics Used: *Elementary Algebra*

In this section we shall describe a simple card trick. The description is divided into three parts. The first part tells what an innocent observer sees. The second tells what the demonstrator does. But the third, a mathematical description, tells why the trick works.

An Observer Describes the Following Trick: The demonstrator forms several piles of cards, face up. It is noted that the piles are uneven. After many observations, it is further noted that a king stands alone as a pile, a queen is covered by one card, a jack by two cards, a ten by three and likewise so that an ace is covered by twelve cards; that is, there are a maximum of thirteen cards in a pile. The piles are turned over and arranged in any order by a viewer, while the demonstrator looks away. Then, another viewer is asked to pick up all but three piles, and the discarded piles are added to the remaining cards in the dealer's hand and shuffled. Next, a viewer is asked to uncover the top card in two of the three remaining piles. This done, the demonstrator counts the cards in hand quietly, and then correctly announces the numerical value of the top card in the remaining pile.

What is the Demonstrator's Secret?   When counting out each pile, the demonstrator starts with the face value of the first card and counts up to thirteen. (An ace counts as 1, a jack as 11, a queen as 12, and a king as 13.) For instance, if the first card is an eight, the count will be 8, 9, 10, 11, 12, 13, giving six cards in the pile. The next card is used to begin a new pile. The process continues until it is clear that there are insufficient cards to make another pile. The extra cards are kept in hand. After the piles are turned over, rearranged, and all but three piles discarded, all cards in hand are shuffled.

The secret is that after the top cards of two piles are exposed, the demonstrator adds ten to the total face value of these cards and counts out that number of cards from those being held. The number of cards remaining in hand is the face value of the unexposed top card of the third pile.

Is this a memorization trick? The above procedure shows it is not, though it may look like one.

Why Does It Work?   If the first card dealt to a pile (this eventually becomes the top card when the piles are turned over) is say 8, then 5 more cards are added to make up to 13. That pile contains 6 cards or $[13 - (8 - 1)]$ cards. Likewise, if the eventual top card is 10, we know that the pile contains 4 cards or $[13 - (10 - 1)]$ cards. In general, if the eventual top card has value x, then the pile contains $[13 - (x - 1)]$ cards.

Therefore, if $a, b,$ and c are the values of the eventual top cards in the three remaining piles, the total number of cards in the three piles would be

$$[13 - (a - 1) + 13 - (b - 1) + 13 - (c - 1)] \quad \text{or} \quad (42 - a - b - c) \text{ cards.}$$

The total number of cards in the dealer's hand would be

$$[52 - (42 - a - b - c)] = 10 + a + b + c.$$

Now, if a and b are the values turned up, then the dealer would count out $10 + a + b$ cards leaving c cards in hand, which is the value of the unexposed top card of the third pile.

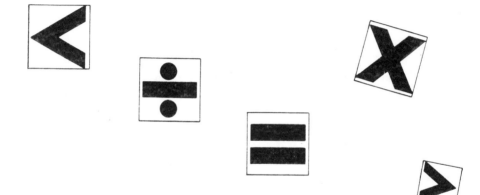

A Fishy Story

Mathematics Used: *Elementary Algebra and Probability*

Consider the problem of estimating the number of fish in a certain lake. One way is the following: Randomly select a spot on the lake and using a net, get a catch of fish. Suppose 200 are caught. These fish are then tagged and returned. One week later the process is repeated. This time 100 are caught and 40 are found to be tagged. This suggests that 40/100 or .4 of the total number of fish in the lake are tagged. If F denotes this total number, then

$$.4F = 200$$
$$F = \frac{200}{.4}$$
$$F = 500.$$

We can generalize the above process as follows:
Let
N_1 = number of fish in first catch (which are tagged)
N_2 = number of fish in second catch
T = number of tagged fish in second catch
F = total number of fish in lake.
Results suggest

$$\frac{T}{N_2}F = N_1$$
$$F = \frac{N_1 N_2}{T}. \tag{1}$$

The result in (1) is called a **basic estimate**.
Example 1: Suppose the following:

$$60 = \text{number of fish in the first catch}$$
$$80 = \text{number of fish in the second catch}$$
$$10 = \text{number of tagged fish in the second catch}.$$

Then a basic estimate is given by

$$F = \frac{60 \cdot 80}{10} = 480.$$

We now examine this estimation problem from a different point of view. Suppose we play the same game with a small goldfish pond. Suppose

$$N_1 = 3$$
$$N_2 = 3$$
$$T = 1.$$

32

It should be clear that there are at least 5 fish in the pond. Our basic estimate would be

$$F = \frac{3 \cdot 3}{1} = 9.$$

We now calculate the probability that one would obtain the given results on the second catch if there were actually 5 fish in the pond.

$$P(\text{results}|5) = \frac{\binom{3}{1}\binom{2}{2}}{\binom{5}{3}} = .3.$$

We can also calculate the probability that one would obtain the given results on the second catch if there were actually 6 fish in the pond.

$$P(\text{results}|6) = \frac{\binom{3}{1}\binom{3}{2}}{\binom{6}{3}} = .45.$$

It is not unreasonable to say that it is "more likely" that there are 6 fish in the pond than 5.

Let us simplify our notation and write $P(n)$ for the probability that one would obtain the given results on the second catch if there were actually n fish in the pond. That is

$$P(n) = P(\text{results}|n)$$

We can then form the following table:

n	$P(n)$
5	.30
6	.450
7	.51429
8	.53571
9	.53571
10	.5250
11	.38182

(The reader should check at least one value of $P(n)$ for n other than 5 or 6.)

From the above table it would appear that the "most likely" number of fish in the pond is 8 or 9. Before accepting this, however, there is one question to answer. Namely, if the table were continued, is it possible that at some point the probabilities would begin to increase again? In order to answer this question we consider

$$\frac{P(K)}{P(K+1)}$$

Note that the probabilities in the table will be increasing when

$$\frac{P(K)}{P(K+1)} < 1$$

and decreasing when

$$\frac{P(K)}{P(K+1)} > 1.$$

Now,

$$P(K) = \frac{\binom{3}{1}\binom{K-3}{2}}{\binom{5}{3}}$$

and

$$P(K+1) = \frac{\binom{3}{1}\binom{(K+1)-3}{2}}{\binom{K+1}{3}}.$$

Hence

$$\frac{P(K)}{P(K+1)} = \frac{\dfrac{\binom{3}{1}\binom{K-3}{2}}{\binom{K}{3}}}{\dfrac{\binom{3}{1}\binom{(K+1)-3}{2}}{\binom{K+1}{3}}}$$

$$= \frac{\binom{K+1}{3}\binom{K-3}{2}}{\binom{K}{3}\binom{K-2}{2}}$$

$$= \frac{\dfrac{(K+1)!}{3!(K-2)!}\cdot\dfrac{(K-3)!}{2!(K-5)!}}{\dfrac{K!}{3!(K-3)!}\,\dfrac{(K-2)!}{2!(K-4)!}}$$

$$= \frac{(K+1)!(K-3)!(K-3)!(K-4)!}{K!(K-2)!(K-2)!(K-5)!}$$

$$= \frac{(K+1)(K-4)}{(K-2)(K-2)}$$

$$= \frac{K^2-3K-4}{K^2-4K+4}.$$

Now

$$\frac{K^2 - 3K - 4}{K^2 - 4K + 4} > 1$$

is equivalent to

$$K^2 - 3K - 4 > K^2 - 4K + 4$$
$$K > 8.$$

That is, past 8 in the table, the probabilities will continue to decrease. Hence we can say that the "most likely" number of fish in the pond is 8 or 9. We call 8 or 9 a *maximum likelihood estimate*. Recall that our basic estimate was 9.

Practice Problem: Suppose $N_1 = 3$, $N_2 = 4$, and $T = 2$. What is a basic estimate for the number of fish in the pond? Compute $P(5), P(6), P(7)$, $P(8)$, and $P(9)$. What is the "most likely" number of fish in the pond? Verify that the probabilities $P(n)$ will decrease after a certain specified value of n.

A Replace Case

Mathematics Used: *Algebra, Inequalities*

A major problem facing many people is the problem of when to replace a piece of equipment (such as an automobile). Suppose a new piece of equipment costs $10,000. The costs for repair and maintenance during each of the following ten years are given below.

Table 1

Year	Cost (in dollars)
1	500
2	780
3	940
4	1100
5	1200
6	1430
7	1920
8	2300
9	3300
10	4400

In order to simplify what follows, we shall assume that there is no trade-in or salvage value for this piece of equipment. Now, if we replace yearly, we incur a yearly cost of $10,500. If we replace every two years, we incur an **average** yearly cost of

$$\frac{10{,}500 + 780}{2} = 5640 \text{ (dollars)}.$$

If we replace every three years, we incur an average yearly cost of

$$\frac{10{,}500 + 780 + 940}{3} = 4073\frac{1}{3} \text{ (dollars)}.$$

Continuing in this fashion, we obtain Table 2.

Table 2 suggests that the "optimal stragegy" would be to replace this piece of equipment every eight years. In making this statement there is one **major** assumption being made and that is the following: If one would continue the calculations of Table 2 for years 11 and 12, etc., **the average yearly cost would continue to rise**. Is this indeed the case or is it possible that at some point the average yearly cost begins to diminish and may even diminish to a figure which is less than $2521.25?

We now use some basic algebra to analyze this problem. Suppose the original cost of a piece of equipment is C (dollars) and that the maintenance and repair costs during years 1, 2, 3, etc. are given, respectively, by a_1, a_2, a_3, etc. If the machine is kept for n years, the average yearly cost is given by

$$\frac{C + a_1 + a_2 + a_3 + \cdots + a_n}{n}.$$

36

Table 2

Year	Maintenance and Repair Costs (in dollars)	Average Yearly Costs (in dollars)
1	500	10,500
2	780	5,640
3	940	$4,073\frac{1}{3}$
4	1100	3,330
5	1200	2,904
6	1430	$2,658\frac{1}{3}$
7	1920	2,552.86
8	2300	2,521.25
9	3300	2,607.78
10	4400	2,787

Now, suppose we encounter the situation where the average yearly cost, if kept k years, is greater than the average cost if kept $k - 1$ years, $k > 1$. That is,

$$\frac{C + a_1 + a_2 + a_3 + \cdots + a_{k-1}}{k - 1} < \frac{C + a_1 + a_2 + a_3 + \cdots + a_k}{k}. \tag{1}$$

Letting

$$A(n) = a_1 + a_2 + \cdots + a_n,$$

we can rewrite (1) as

$$\frac{C + A(k - 1)}{k - 1} < \frac{C + A(k)}{k}. \tag{2}$$

We now wish to establish the conditions under which the average yearly cost will continue to rise if this piece of equipment is kept for one more year (i.e., $k + 1$) years. Hence, we must find conditions under which

$$\frac{C + A(k)}{k} < \frac{C + A(k + 1)}{k + 1} \tag{3}$$

$$C\left[\frac{1}{k} - \frac{1}{k + 1}\right] < \frac{A(k + 1)}{k + 1} - \frac{A(k)}{k}$$

$$C\left[\frac{1}{k(k + 1)}\right] < \frac{kA(k + 1) - (k + 1)A(k)}{k(k + 1)}$$

$$C < kA(k + 1) - (k + 1)A(k). \tag{4}$$

Recalling that $A(k + 1) = A(k) + a_{k+1}$, we may write (4) as

$$C < k[A(k) + a_{k+1}] - (k + 1)A(k)$$

$$C < ka_{k+1} - A(k). \tag{5}$$

37

We know from our assumption that the average yearly cost if kept k years is greater than the average yearly cost if kept $k - 1$ years, $k > 1$, that (5) is true if k is replaced by $k - 1$ so that

$$C < (k-1)a_k - A(k-1).$$

It follows that (5) will hold if one can show

$$(k-1)a_k - A(k-1) < ka_{k+1} - A(k). \tag{6}$$

Recalling that $A(k) = A(k-1) + a_k$ one can write (6) as

$$(k-1)a_k - A(k-1) < ka_{k+1} - A(k-1) - a_k$$
$$ka_k < ka_{k+1}$$
$$a_k < a_{k+1}. \tag{7}$$

Now, (7) will hold as long as maintenance and repair costs continue to rise. From this result one can conclude the following: Suppose in a table, such as Table 2, one encounters the situation where the average yearly cost increases from one year to the next. **As long as the maintenance and repair costs continue to increase, the average yearly cost will also continue to increase.** Hence, under such conditions one cannot be misled by a list of figures such as those given in Table 2.

Practice Problem: When should the piece of equipment, described in this section, be replaced if the initial cost is $8,000?

The Mouse That Roared

Mathematics Used: *Algebra, Logarithms*

A single foreign cell is injected into a mouse. One day later, there are 4 cells, two days later, 16 cells, etc., as given in the table.

Day	# Cells
Now	1
1	4
2	16
3	64
4	256
.	.
.	.
.	.
.	.
T	4^T (letting $T = 0, 1, 2, 3, \ldots$)

Suppose that the mouse will die if the number of cells exceeds 1,000,000. Our problem is the following: A treatment is available which is 96% effective (i.e., kills 96% of the cells). When **must** the treatment be given in order to keep the mouse alive?

Now, one finds that (use a calculator)

$$4^9 = 262,144$$

while

$$4^{10} = 1,048,576.$$

Thus, the first treatment must be given between day 9 and day 10.

To be more precise one might
i) use "linear interpolation".

Power	Number
10	1,048,576
x	1,000,000
9	262,144

$$\frac{x-9}{10-9} = \frac{737,856}{786,432}$$
$$x - 9 = .938232$$
$$x = 9.938232 \text{ (days)}.$$

(This is slightly more than 9 days, 22 1/2 hours.)

ii) Solve the equation $4^T = 1,000,000$ for T.

$$4^T = 10^6$$
$$T \log 4 = 6 \log 10$$
$$T = \frac{6 \log 10}{\log 4}$$
$$T = 9.965784 \text{ (days)}.$$

(This is slightly more than 9 days, 23 hours.)

Suppose that the number reaches 1,000,000 when the first treatment is given. There are then 40,000 cells left in the mouse. One day later, there are 4(40,000) cells, etc.

Day	# Cells
Now	40,000
1	4(40,000)= 160,000
2	16(40,000)= 640,000
3	64(40,000)= 2,560,000
.	.
.	.
.	.
T	$4^T(40,000)$

It is clear that the next treatment must be given between day 2 and day 3. We solve the following:

$$4^T(40,000) = 1,000,000$$
$$4^T = 25$$
$$T \log 4 = \log 25$$
$$T = \frac{\log 25}{\log 4}$$
$$T = 2.321928.$$

This is slightly more than 2 days, 7.72 hours.

Practice Problem: A single foreign cell is injected into a mouse. On each day that follows, the number of foreign cells is double the number present on the previous day. If 600,000 foreign cells will kill this mouse, when **must** the treatment (described in this section) be given?

The Child Prodigy

Mathematics Used: *Algebra, Elementary Probability*

Many problems facing business people are similar to those facing the young-ster in the now classic "newspaper carrier problem."

The carrier buys newspapers for 5¢ and sells them for 10¢. S/he is given 3¢ the following day for each copy which is not sold.

The carrier decides to try to predict how many papers s/he is going to sell to maximize long-term profit. After studying what happens over a 100-day period and taking into account the demand (not just the number sold), s/he compiles Table 1 and derives probabilities from this table. For example, the probability of the demand being 4 papers per day is taken to be 1/100, the probability of a demand of 17 is 3/100, and probability of a demand of 26 is 6/100, etc. Using these probabilities, the carrier computes expected revenue for each of the numbers 0 through 36. For example, if s/he buys 6 papers, observes, from the cumulative column, that the probability of selling 5 or less papers is 7/100 so that the probability of selling all 6 papers is 93/100. Thus, expected profit is given by

$$(.93)(.30) + (.03)(.23) + (.01)(.16) + (.02)(.09) + (.01)(-.05) = .2888.$$

As another example, if s/he buys 10 papers, observe that the probability of selling all 10 papers is 88/100. Expected profit in this case is thus given by

$$(.88)(.50) + (.01)(.43) + (.01)(.36)$$
$$+ (.01)(.29) + (.02)(.22) + (.03)(.15)$$
$$+ (.01)(.08) + (.02)(.01) + (.01)(-.13) = .4594.$$

After much time with these tedious computations, the carrier finds that s/he should purchase 28 papers to maximize expected profit.

The carrier could have saved considerable time by examining the following: Let x denote an integer, $0 \leq x \leq 36$. Let $P(x)$ denote the probability of selling x or less papers. These probabilities may be read directly from the "cumulative" column. For example, $P(10) = 15/100$, $P(17) = 31/100$, etc.

Now suppose s/he orders x papers and considers what would happen if **one more** paper (i.e., $x+1$ papers) was ordered. On the additional paper s/he would make .05 with probability $1 - P(x)$ and would lose .02 with probability $P(x)$. Thus, expected profit on the additional paper would be

$$(.05)(1 - P(x)) + (-.02)P(x)$$

or

$$.05 - .07P(x).$$

An additional paper should be purchased if

$$.05 - .07P(x) \geq 0$$
$$P(x) \leq \frac{.05}{.07} = \frac{5}{7}.$$

Table 1

Demand	# of Days	Cumulative # of Days
0	0	0
1	1	1
2	0	1
3	2	3
4	1	4
5	3	7
6	2	9
7	1	10
8	1	11
9	1	12
10	3	15
11	1	16
12	2	18
13	4	22
14	1	23
15	3	26
16	2	28
17	3	31
18	4	35
19	4	39
20	4	43
21	6	49
22	2	51
23	4	55
24	3	58
25	4	62
26	6	68
27	3	71
28	7	78
29	5	83
30	4	87
31	3	90
32	4	94
33	4	98
34	1	99
35	0	99
36	1	100

Thus, the carrier should purchase an additional copy as long as

$$P(x) \leq \frac{5}{7}.$$

Now, $5/7 = .714$ (to three decimal places). From the table, we see that $P(27) = .71$, while $P(28) = .78$. Thus,

$$P(27) < \frac{5}{7} < P(28)$$

and it follows that the "optimal" purchase is 28 papers.

Practice Problem: Suppose the newspaper company changes its policy and gives only 2¢ the following day for each copy which is not sold. What is the "optimal" purchase under these conditions?

That #%&*?/ Mortgage Payment

Mathematics Used: *Algebra, Geometric Series*

Suppose that an individual obtains an $80,000 mortgage at an interest rate of 9% for a period of 20 years. Each month, the individual is to make a payment of P dollars (the same each month). We show here how to determine the number P.

At the end of the first month, the individual pays P dollars (the first payment). A part of this, say p_1, is on the principal, while $\frac{.09}{12}(80,000) = .0075(80,000) = 600$ dollars is for interest. We thus see that $P > \$600$. We write

$$P = \underbrace{p_1}_{\text{on principal}} + \underbrace{(80,000).0075.}_{\text{on interest}} \tag{1}$$

In order to simplify what follows, let $i = .0075$, so that (1) becomes

$$P = p_1 + (80,000)i. \tag{2}$$

At the end of the second month, the payment P is p_2 dollars on the principal plus interest on the remaining principal $80,000 - p_1$. Thus,

$$P = p_2 + (80,000 - p_1)i. \tag{3}$$

Continuing, at the end of the third month, we have

$$P = p_3 + (80,000 - p_1 - p_2)i \tag{4}$$

and at the end of the fourth month

$$P = p_4 + (80,000 - p_1 - p_2 - p_3)i. \tag{5}$$

In general, at the end of month k, we have

$$P = p_k + (80,000 - p_1 - p_2 - \cdots - p_{k-1})i. \tag{6}$$

Now, from (2) and (3), we have

$$p_1 + (80,000)i = p_2 + (80,000 - p_1)i$$

or

$$p_2 = p_1(1 + i). \tag{7}$$

From (3) and (4), we have

$$p_2 + (80,000 - p_1)i = p_3 + (80,000) - p_1 - p_2)i$$

or

$$p_3 = p_2(1 + i)$$

44

which, together with (7), yields

$$p_3 = p_1(1+i)^2. \tag{8}$$

From (4) and (5), we have

$$p_4 = p_3(1+i)$$

which, together with (8), yields

$$p_4 = p_1(1+i)^3. \tag{9}$$

Summarizing, we see

$$\left.\begin{array}{l} p_1 \\ p_2 = p_1(1+i) \\ p_3 = p_1(1+i)^2 \\ p_4 = p_1(1+i)^3 \end{array}\right\} \tag{10}$$

This pattern continues (the student can show this by induction) so that

$$p_k = p_1(1+i)^{k-1}.$$

Now, p_1, p_2, p_3, etc. are payments on the principal and so

$$p_1 + p_2 + p_3 + \cdots + p_{240} = 80,000$$

or using the summary in (10) and the general term, we have

$$p_1 + p_1(1+i) + p_1(1+i)^2 + \cdots + p_1(1+i)^{239} = 80,000.$$

The left-hand side of the above equation is a geometric series. Using the formula for the sum of such series, we have

$$\frac{p_1[(1+i)^{240} - 1]}{(1+i) - 1} = 80,000$$

or

$$p_1 = \frac{(80,000)i}{(1+i)^{240} - 1}.$$

Using interest tables, logarithms, or a calculator, we obtain

$$p_1 = 119.78 \text{ (to the nearest cent)}.$$

So that (from (1)), we have

$$P = 119.78 + 600$$

or

$$P = \$719.78.$$

Note that the total amount paid over the 20 year period will be $172,747.20.

Practice Problem: Determine the monthly payments on a $70,000 mortgage at an interest rate of 10% for a period of 25 years.

A Smug Drug

Mathematics Used: *Algebra, Geometric Sequences and Series*

A doctor determines that 10 units of a certain drug are now in a body. At the end of any hour, assuming no injections of the drug are given, there is one-third of the active concentrate that existed in the body at the beginning of the hour. The rest is removed as waste or made inactive by chemical reactions in the body. Let us now chart the amount in the body:

Table 1

Hour	Active Concentrate
0	10
1	$(1/3)10$
2	$(1/3)^2 10$
3	$(1/3)^3 10$
4	$(1/3)^4 10$
.	.
.	.
.	.
24	$(1/3)^{24} 10$

Suppose that after 24 hours, an injection of 10 units is given. Immediately **after** this first injection, the amount A_1 of active concentrate is given by

$$A_1 = 10 + (1/3)^{24} 10.$$

Writing B_1 for the amount of active concentrate immediately **before** this first injection, we see from Table 1 that

$$B_1 = (1/3)^{24} 10.$$

Now, beginning with A_1 after the first injection, one hour later we have $\frac{1}{3} A_1$, 2 hours later $(1/3)^2 A_1$, and 3 hours later $(1/3)^3 A_1$, etc. Thus, if a second injection is to be given 24 hours after the first, the amount B_2 of active concentrate in the body **before** the second injection is given by

$$B_2 = (1/3)^{24} A_1$$

or

$$B_2 = (1/3)^{24} 10 + (1/3)^{2 \cdot 24} 10.$$

Letting A_2 denote the amount of active concentrate immediately after the second injection, we have

$$A_2 = 10 + (1/3)^{24} 10 + (1/3)^{2 \cdot 24} 10.$$

Now, if a third injection is given **24** hours after the second, and if B_3 and A_3 denote the amount of active concentrate in the body **before** and **after** this injection, then (the student should work through the argument)

$$B_3 = (1/3)^{24}10 + (1/3)^{2 \cdot 24}10 + (1/3)^{3 \cdot 24}10$$

and

$$A_3 = 10 + (1/3)^{24}10 + (1/3)^{2 \cdot 24}10 + (1/3)^{3 \cdot 24}10.$$

Similarly, if a fourth injection is given **24** hours after the third, and if B_4 and A_4 denote the amount of active concentrate in the body **before** and **after** this injection, then (the student should work through the argument)

$$B_4 = (1/3)^{24}10 + (1/3)^{2 \cdot 24}10 + (1/3)^{3 \cdot 24}10 + (1/3)^{4 \cdot 24}10$$

and

$$A_4 = 10 + (1/3)^{24}10 + (1/3)^{2 \cdot 24}10 + (1/3)^{3 \cdot 24}10 + (1/3)^{4 \cdot 24}10.$$

Summarizing, we see the following:

$$B_1 = (1/3)^{24}10$$
$$B_2 = (1/3)^{24}10 + (1/3)^{2 \cdot 24}10$$
$$B_3 = (1/3)^{24}10 + (1/3)^{2 \cdot 24}10 + (1/3)^{3 \cdot 24}10$$
$$B_4 = (1/3)^{24}10 + (1/3)^{2 \cdot 24}10 + (1/3)^{3 \cdot 24}10 + (1/3)^{4 \cdot 24}10$$

and

$$A_1 = 10 + (1/3)^{24}10$$
$$A_2 = 10 + (1/3)^{24}10 + (1/3)^{2 \cdot 24}10$$
$$A_3 = 10 + (1/3)^{24}10 + (1/3)^{2 \cdot 24}10 + (1/3)^{3 \cdot 24}10$$
$$A_4 = 10 + (1/3)^{24}10 + (1/3)^{2 \cdot 24}10 + (1/3)^{3 \cdot 24}10 + (1/3)^{4 \cdot 24}10.$$

The pattern should now be clear. Namely, if the process continues and if B_k and A_k denote the amounts of active concentrate before and after the kth injections, then

$$B_k = (1/3)^{24}10 + (1/3)^{2 \cdot 24}10 + (1/3)^{3 \cdot 24}10 + \cdots + (1/3)^{k \cdot 24}10$$

and

$$A_k = 10 + (1/3)^{24}10 + (1/3)^{2 \cdot 24}10 + (1/3)^{3 \cdot 24}10 + \cdots + (1/3)^{k \cdot 24}10.$$

Now, B_k and A_k are given by geometric series. Using the formula for the sum of each series, we have

$$B_k = \frac{(1/3)^{24}10(1 - (1/3)^{k \cdot 24})}{1 - (1/3)^{24}}$$

and

$$A_k = \frac{10(1 - (1/3)^{(k+1)\cdot 24})}{1 - (1/3)^{24}}.$$

As k gets large, $(1/3)^{k\cdot 24}$ and $(1/3)^{(k+1)\cdot 24}$ get close to zero. Hence, B_k and A_k get close, respectively, to B and A given by

$$B = \frac{(1/3)^{24}10}{1 - (1/3)^{24}}$$

and

$$A = \frac{10}{1 - (1/3)^{24}}.$$

Suppose, at some point in time, that the active concentrate in the body before an injection actually reaches B. Then immediately after the injection, there are

$$10 + B = 10 + \frac{(1/3)^{24}10}{1 - (1/3)^{24}} = \frac{10}{1 - (1/3)^{24}} = A$$

units of active concentrate in the body. Before the next injection 24 hours later, there are

$$(1/3)^{24}A = \frac{(1/3)^{24}10}{1 - (1/3)^{24}} = B$$

units of active concentrate in the body.

Thus, **for all practical purposes**, the active concentrate behaves according to the following graph:

concentrate

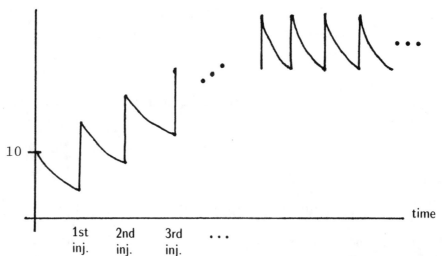

10

time

1st   2nd   3rd $\cdots$
inj.   inj.   inj.

A problem which has been mentioned by a medical doctor is that for drugs which behave in this manner and which are used in the treatment of diseases,

there is a good chance that the body will become immune to the drug when (and if) the drug behaves in the uniform manner given by

Practice Problems 1: Suppose that 6 units of a certain drug are now in a body. At the end of any hour, assuming no injections of the drug are given, there is one-half of the active concentrate that existed in the body at the beginning of the hour. Suppose injections of 6 units are given every 9 hours. Let B_k and A_k denote the amount of active concentrate in the body **before** and **after**, respectively, the kth injection. Use the formula for the sum of a geometric series to find compact expressions for B_k and A_k. Find A and B. Show that if the active concentrate in the body reaches B before an injection, then immediately after the injection there are A units in the body. Show that before the next injection, there are B units of active concentrate in the body.

Practice Problem 2: Generalize the problem of this section in the following way: Suppose that n units of a certain drug are now in a body. At the end of any hour, assuming no injections of the drug are given, there is a part r, $0 < r < 1$, of the active concentrate that existed in the body at the beginning of the hour. Suppose injections of n units are given every t hours. Let B_k and A_k denote the amount of active concentrate in the body **before** and **after**, respectively, the kth injection. Use the formula for the sum of a geometric series to find compact expressions for B_k and A_k. Find A and B. Show that if the active concentrate in the body reaches B before an injection, then immediately after the injection there are A units in the body. Show that before the next injection, there are B units of active concentrate in the body.

An Expected Die

Mathematics Used: *Elementary Probability*

Would you play the following game? A die is tossed 60 times. On each toss, if a 1 shows, you pay me $3. If a 2 or 3 shows, I pay you $12, and if a 4, 5, or 6 shows, you pay me $6.

When a class of college students was asked this question, approximately half of the students said they would be willing to play the game, while the other half said they were not willing to play (in some cases for non-monetary reasons).

The students who were willing to play gave little or no reason for their willingness. All of the students, however, agreed to the following: During the 60 throws of the die, one would **expect** a 1 to appear on 10 throws, a 2 or 3 on 20 throws, and a 4, 5, or 6 on 30 throws. The **expected monetary return** (to you) would be

$$10(-3) + 20(12) + 30(-6).$$

A negative number such as the (-3) indicates a payment (gain) to me, but a loss to you. A positive number such as the (12) indicates a payment (gain) to you, but a loss to me. Now,

$$10(-3) + 20(12) + 30(-6) = \$30.$$

That is, **you** would **expect** to gain $30 in the 60 throws of the die. This amounts to an **average** gain, per throw, of $.50.

Under the same payoff rules, suppose that the die is tossed 90 times. Then one would **expect** a 1 to appear on 15 throws, a 2 or 3 to show on 30 throws, and a 4, 5, or 6 to show on 45 throws. Thus, the expected monetary returns (to you) would be

$$15(-3) + 30(12) + 45(-6) = \$45.$$

Again, this amounts to an **average** gain of $.50 per throw.

In both cases, we obtained an average gain (to you) of $.50 per throw. This is not a coincidence. We could have obtained this number in the following way: On any particular throw of the die, the probability of a 1 showing is 1/6, of a 2 or 3 showing is 2/6, and of a 4, 5, or 6 showing is 3/6. We form the sum

$$\frac{1}{6}(-3) + \frac{2}{6}(12) + \frac{3}{6}(-6) = .50.$$

That is, we simply form the sum of products obtained by multiplying probabilities and associated gain (or loss). In this way we obtain the average gain per throw. For example, in 150 throws of the die, one would **expect** to gain $75.

Let us return to the 60 tosses of a die where a 1 was **expected** 10 times, a 2 or 3 was **expected** 20 times, and a 4, 5, or 6 was **expected** 30 times. We know that it is quite unlikely that the results will be exactly as **expected**. One could, of course, perform an experiment and actually toss a die 60 times and count the results in each category. An easier way is to have a computer simulate

the toss of a die. For this purpose the following BASIC program was written for an APPLE microcomputer:

PROGRAM

```
10   PRINT "WHAT IS N?"
20   INPUT N
30   I = 1
50   X1 = 0
60   X2 = 0
70   X3 = 0
80   X4 = 0
90   X5 = 0
100  X6 = 0
110  VI = RND(1) * 6 + 1
120  IF INT(VI) = 1 THEN X1 = X1 + 1
130  IF INT(VI) = 2 THEN X2 = X2 + 1
140  IF INT(VI) = 3 THEN X3 = X3 + 1
150  IF INT(VI) = 4 THEN X4 = X4 + 1
160  IF INT(VI) = 5 THEN X5 = X5 + 1
170  IF INT(VI) = 6 THEN X6 = X6 + 1
180  I = I + 1
190  IF I <= N GOTO 110
200  PRINT X1, X2, X3, X4, X5, X6
210  END
```

INSTRUCTIONS FOR PROGRAM

When one types RUN the question "WHAT IS N?" will appear. The number N represents the number of times you wish the computer to toss a die. If you type 60, for example, the computer will respond by giving you numbers such as the following:

$$9 \quad 12 \quad 8 \quad 7 \quad 11 \quad 13$$

These numbers represent the following:

9	–	No. of 1's
12	–	No. of 2's
8	–	No. of 3's
7	–	No. of 4's
11	–	No. of 5's
13	–	No. of 6's

We ran this program for $N = 60$ tosses a number of times and have listed below the results of two of these runs:

Run I:

1	2	3	4	5	6
9	9	5	13	14	10

Grouping results gives the following:

$$1 \} \quad 9 \text{ times}$$

$$\left.\begin{matrix} 2 \\ 3 \end{matrix}\right\} \quad 14 \text{ times}$$

$$\left.\begin{matrix} 4 \\ 5 \\ 6 \end{matrix}\right\} \quad 37 \text{ times}$$

Assuming the same payoff rules as before, your actual gain is $9(-3)+14(12)+37(-6) = -\$81$, whereas your **expected** gain is \$30.

Run II:

1	2	3	4	5	6
12	9	11	8	5	15

Grouping results gives the following:

$$1 \} \quad 12 \text{ times}$$

$$\left.\begin{matrix} 2 \\ 3 \end{matrix}\right\} \quad 20 \text{ times}$$

$$\left.\begin{matrix} 4 \\ 5 \\ 6 \end{matrix}\right\} \quad 28 \text{ times}$$

Your actual gain is $12(-3)+20(12)+28(-6) = \$36$, whereas your **expected** gain is \$30.

Suppose we have the computer toss a die 1200 times. Your expected gain is $1200(.50) = \$600$.

A computer run resulted in the following:

1	2	3	4	5	6
179	204	229	183	217	188

When grouped, the results are as follows:

$$1 \} \quad 179 \text{ times}$$

$$\left.\begin{matrix} 2 \\ 3 \end{matrix}\right\} \quad 433 \text{ times}$$

$$\left.\begin{matrix} 4 \\ 5 \\ 6 \end{matrix}\right\} \quad 588 \text{ times}$$

Your actual gain is $179(-3) + 433(12) + 588(-6) = \$1131$.

Play a Sane Game

Mathematics Used: *Algebra, Elementary Probability*
(Review Fantastik 17)

Let us now consider the following game. Two players A and B sit across a table from each other. Each has a coin. For 100 times, the coins are going to be placed simultaneously on the table with the following payoff rules (T for "tail" and H for "head"):

A shows	B shows	
H	H	B pays A \$3
H	T	A pays B \$1
T	H	A pays B \$6
T	T	B pays A \$4

It is standard practice to put such a game in "matrix" form. That is, we tabulate the above information as a two by two array given as follows:

$$\begin{array}{cc} & B \\ & \begin{matrix} H & \quad T \end{matrix} \\ A \begin{matrix} H \\ T \end{matrix} & \begin{pmatrix} 3 & -1 \\ -6 & 4 \end{pmatrix} \end{array}$$

All of the entries in this array represent payments (or losses, as the case may be) to A. For example, the 3 in the first row, first column, indicates the payment of \$3 to A if both show heads, while the -1 in the first row, second column, indicates a payment (loss) of $-\$1$ to A which means that A pays \$1 to player B.

Let us suppose that player A decides to show heads (H) 1/4 of the time and tails (T) 3/4 of the time, while player B decides to show heads 1/2 of the time and tails 1/2 of the time. This information is given by the following array:

$$\begin{array}{cc} & B \\ & \begin{matrix} \frac{1}{2} & \quad \frac{1}{2} \\ H & \quad T \end{matrix} \\ A \begin{matrix} \frac{1}{4}H \\ \frac{3}{4}T \end{matrix} & \begin{pmatrix} 3 & -1 \\ -6 & 4 \end{pmatrix} \end{array}$$

Now, on $1/4 \cdot 1/2 = 1/8$ of the time, we can expect both to show heads, while on $3/4 \cdot 1/2 = 3/8$ of the time, we can expect A to show tails and B to show heads. Continuing in this manner, we can find the **average** gain, per show, to player A by computing

$$\frac{1}{8}(3) + \frac{1}{8}(-1) + \frac{3}{8}(-6) + \frac{3}{8}(4) = -\frac{4}{8} = -.50.$$

That is, player A can **expect** to lose $\$.50$ per show.

A major problem of game theory is to determine what part of the time player A should show heads and what part of the time player A should show tails in order to form an **optimal strategy**.

Let us suppose that player A decides to show heads "p" of the time, p a number $0 \leq p \leq 1$. Then tails will show $1 - p$ of the time (in our previous example, $p = 1/4$ and $1 - p = 3/4$). Also, let us suppose that player B decides to show heads "q" of the time and tails $1 - q$ of the time. We indicate these decisions in the following array:

$$
\begin{array}{cc}
 & B \\
 & \begin{array}{cc} q & 1-q \\ H & T \end{array} \\
A\;\; \begin{array}{cc} p & H \\ 1-p & T \end{array} & \begin{pmatrix} 3 & -1 \\ -6 & 4 \end{pmatrix}
\end{array}
$$

The expected gain G to player A is given by

$$G = pq(3) + p(1 - q)(-1) + (1 - p)q(-6) + (1 - p)(1 - q)(4)$$

or, expanding and collecting terms,

$$G = 14pq - 5p - 10q + 4$$
$$G = 14\left(pq - \frac{5}{14}p - \frac{10}{14}q\right) + 4$$
$$G = 14\left(p - \frac{10}{14}\right)\left(q - \frac{5}{14}\right) + \frac{6}{14}.$$

From this we see the following: If player A shows heads $10/14$ (i.e., $p = 10/14$) of the time, then **no matter what player B does**, player A can expect to gain an average of $\$6/14$ per show. If player A would take $p > 10/14$ (so that $p - 10/14$ is positive) and player B became aware of this, then player B could take $q < 5/14$ (so that $q - 5/14$ is negative) and detract from the gain of $\$6/14$ and possibly even force a loss on player A by making G negative (the student should show that this is indeed a possibility). Thus, we say that the **optimal strategy** for A is to take $p = 10/14$. Since $10/14 = 5/7$, one way for player A to accomplish this would be to divide a circle in 7 equal parts and number the sectors one through seven (see diagram on the next page). Using a spinner, if the arrow falls on 1, 2, 3, 4, or 5, A will show a head, otherwise a tail.

Let us now consider again the equation

$$G = 14\left(p - \frac{10}{14}\right)\left(q - \frac{5}{14}\right) + \frac{6}{14}$$

and examine this result from the point of view of player B. Player B is in trouble, for observe that if player A chooses $p = 10/14$, then no matter what player B does, s/he can expect to lose an average of $\$6/14$ per show. Now, if player B

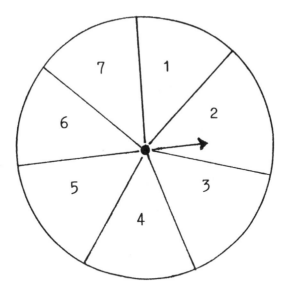

chooses $q > 5/14$ (so that $q - 5/14$ is positive) and player A becomes aware of this decision, then player A could choose $p > 10/14$ (so that $p - 10/14$ is also positive) and player B can expect to lose more than \$6/14 per show. Similarly, if player B chooses $q < 5/14$ (so that $q - 5/14$ is negative) and player A becomes aware of this decision, then player A could choose $p < 10/14$ (so that $p - 10/14$ is also negative) and player B can again expect to lose more than \$6/14 per show. We thus say that the optimal strategy for player B is to choose $q = 5/14$.

Practice Problem 1: Analyze, from the point of view of player A, then player B, each of the following games:

a)

$$B$$
$$A\begin{pmatrix} 2 & -3 \\ -4 & 6 \end{pmatrix}$$

b)

$$B$$
$$A\begin{pmatrix} 3 & -4 \\ -6 & 7 \end{pmatrix}$$

c)

$$B$$
$$A\begin{pmatrix} 6 & 1 \\ -3 & -7 \end{pmatrix}$$

Practice Problem 2: Suppose that there are two possible games being played as follows:

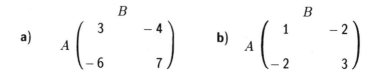

Player A thinks there is a 3/10 chance that game **a**) is being played and a 7/10 chance that game **b**) is being played. Player B is certain that game **b**) is being played. What strategy would you use if you were player A? Player B? (This type of problem was actually encountered when, in 1968, a group of mathematicians studied the following problem: Should we (the U.S.) negotiate a disarmament treaty with the Russians?)

Satellite Hite Site

Mathematics Used: *Trigonometry*

From two tracking stations d miles apart, the elevation angle of a satellite is determined to be θ_1 and θ_2, respectively, where $0 < \theta_1 < 90°$ and $0 < \theta_2 < 90°$, $\theta_2 > \theta_1$, as illustrated in the following figure:

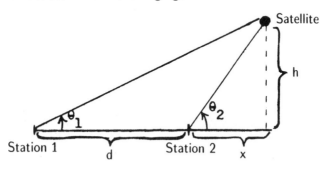

Our problem is to determine h.
We observe that

$$\tan \theta_2 = \frac{h}{x} \qquad \text{or} \qquad x = h \cot \theta_2 \tag{1}$$

and

$$\tan \theta_1 = \frac{h}{d + x}$$

or

$$d + x = h \cot \theta_1 \qquad \text{or} \qquad x = h \cot \theta_1 - d. \tag{2}$$

From (1) and (2), we have

$$h \cot \theta_2 = h \cot \theta_1 - d$$

or

$$h = \frac{d}{\cot \theta_1 - \cot \theta_2}. \tag{3}$$

For example, if $\theta_1 = 28°$, $\theta_2 = 67°$, and $d = 1000$ miles, then

$$h = \frac{1000}{\cot 28° - \cot 67°} = 686.69 \text{ miles.}$$

Practice Problem: Find a formula for h, such as that given in (3), for all other relative positions of the tracking stations and the satellite.

Pray Prey

Mathematics Used: *Algebra, Factoring, Inequalities*

The interaction of mathematics with the real world is frequently the result of four steps:

1) A real world problem is presented and information relevant to that problem is gathered.

2) From the information gathered in step (1), a mathematical model (equation, formula, etc.) is built.

3) Mathematical manipulations are performed on the model. These manipulations follow accepted rules of logic and are based on results which appear in mathematical literature.

4) From the mathematical results of step (3), one now returns to the real world and considers what these mathematical results have to say about the real world problem originally presented in step (1). In some cases, one might use the mathematical results to predict a future state of the world. It is in such cases that the study of statistics interacts with mathematics and the real world.

The following diagram is an aid in understanding the above four steps:

Mathematical Model   →   Mathematical Results

↑   ↓

Real World Problem   Real World

Let us consider the following example which illustrates this process.

1. Real World Problem

An insecticide is sprayed over a large area of land. The insecticide is being used to kill a certain type of insect which is prey for a second type of insect, its predator. Suppose that the insecticide kills a certain number of both prey and predator. Our problem is the following: What ecological changes might occur as a result of this activity?

A mathematician by the name of Volterra considered this type of problem around 1930. Before constructing a mathematical model, it was necessary to gather information concerning this real world problem. In certain cases, available evidence suggested that in the absence of the predator, the prey population would **increase** at a rate proportional to the prey population. The presence of the predator population, however, caused a decrease in the prey population and the predator population. The study of such cases also suggested that in the absence of the prey population (no food), the predator population would decrease (die) at a rate proportional to the predator population. The presence of the prey population (food), however, caused an increase in the predator population which was proportional to the product of the prey population and the predator population.

2. Mathematical Model

At time t we have two populations present, $H(t)$ and $P(t)$, where $H(t)$ is the prey population and $P(t)$ is the predator population. Let $r_H(t)$ denote the

rate at which the prey population is changing at time t, and $r_P(t)$ denote the rate at which the predator population is changing at time t. For example, if t is measured in days and at a certain time t, $r_H(t) = 20$, then at that time the prey population is **increasing** at the rate of 20 per day. However, if $r_H(t) = -20$, then at that time, the prey population is **decreasing** at the rate of 20 per day.

From the information gathered in step (1), we write

$$\left. \begin{array}{l} r_H(t) = kH(t) - lH(t)P(t) \\ r_P(t) = mH(t)P(t) - nP(t) \end{array} \right\}$$

or, more simply,

$$\left. \begin{array}{l} r_H = kH - lHP \\ r_P = mHP - nP \end{array} \right\} \tag{1}$$

where k, l, m, and n are positive numbers.

The two populations are said to be in **natural equilibrium** when $r_H = 0$ and $r_P = 0$. In order for this to occur, we must have, from (1),

$$\left. \begin{array}{l} kH - lHP = 0 \\ mHP - nP = 0 \end{array} \right\} \tag{2}$$

which yields

$$\left. \begin{array}{l} P = \dfrac{k}{l} \\ H = \dfrac{n}{m} \end{array} \right\} .$$

We call these numbers the **equilibrium populations** and denote them by P_0 and H_0. Thus,

$$\left. \begin{array}{l} P_0 = \dfrac{k}{l} \\ H_0 = \dfrac{n}{m} \end{array} \right\} . \tag{3}$$

Suppose the two populations are in natural equilibrium with populations given by (3). Now introduce an agent (such as an insecticide) which kills h (> 0) of the prey and p (> 0) of the predators. We assume that the agent does not annihilate either population. Immediately after the introduction of the agent, the resulting populations are

$$\left. \begin{array}{l} H_0 - h \;\; (> 0) \text{ prey} \\ P_0 - p \;\; (> 0) \text{ predator} \end{array} \right\} . \tag{4}$$

3. Mathematical Results

We now wish to use mathematical manipulations to consider r_H and r_P immediately after the introduction of the agent mentioned above. The populations are now given in (4). From (1), we have

$$\left. \begin{array}{l} r_H = k(H_0 - h) - l(H_0 - h)(P_0 - p) \\ r_P = m(H_0 - h)(P_0 - p) - n(P_0 - p) \end{array} \right\} . \tag{5}$$

Now,

$$r_H = k(H_0 - h) - l(H_0 - h)(P_0 - p)$$
$$= (H_0 - h)[k - l(P_0 - p)]$$
$$= (H_0 - h)\left[k - l\left(\frac{k}{l} - p\right)\right]$$
$$= (H_0 - h)[lp].$$

Since $(H_0 - h) > 0$, $l > 0$, and $p > 0$, it follows that $r_H > 0$. Also,

$$r_P = m(H_0 - h)(P_0 - p) - n(P_0 - p)$$
$$= (P_0 - p)[m(H_0 - h) - n]$$
$$= (P_0 - p)\left[m\left(\frac{n}{m} - h\right) - n\right]$$
$$= (P_0 - p)[-mh].$$

Since $(P_0 - p) > 0$, $m > 0$, and $h > 0$, it follows that $r_P < 0$.

Our mathematical results are thus $r_H > 0$ and $r_P < 0$.

4. Real World

We now wish to interpret the mathematical results $r_H > 0$ and $r_P < 0$. We assumed that the prey and predator populations were initially in natural equilibrium. An agent was then introduced which killed some (but not all) of the prey population and some (but not all) of the predator population. After the introduction of this agent, we have $r_H > 0$ and $r_P < 0$ which tells us that the prey population is now **increasing**, whereas the predator population is now **decreasing**. This is the ecological phenomenon which is set in motion **whenever the assumptions of our model are satisfied**. The ecological problem which presents itself is the following: The predator population, which is now decreasing, can become so small that it is in jeopardy of being annihilated by other causes: e.g., flood, draught, etc. Also, the predator population can become "thin" or "sparse" in which case the chances of reproduction are minimized and can cause a further decrease in the predator population.

The behavior discussed here has been observed. The following is taken from *The Biology of Populations* by MacArthur and Connell.

> Volterra's principle implies that the application of the insecticides will, unless **extremely** destructive to the prey, increase the populations of those insects that are kept in control by other predatory insects. A remarkable confirmation came from the cottony cushion scale insect (*Icerya purchasi*), which, when accidentally introduced from Australia in 1868, threatened to destroy the American citrus industry. Thereupon, its natural Australian predator, a lady-bird beetle, *Novius cardinalis*, was introduced and took hold immediately, reducing the scale to a low level. When DDT was discovered to kill scale insects, it was applied by the orchardists in the hope of further reducing the scale insects. However, in agreement with Volterra's principle, the effect was an increase of the scale insect! This shows the danger of tampering with those aspects of nature that are not understood. (Elton, 1958)

A Locatshun Problum

Mathematics Used: *Algebra, Absolute Value, Graphing*

Let us suppose that there are four machines m_1, m_2, m_3, and m_4 in a certain factory which are located along an aisle (the x-axis) as illustrated in Figure 1.

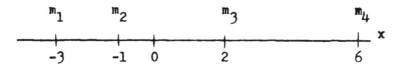

Figure 1

A new machine is to be located somewhere along this aisle. There will be interaction between the existing machines and the new machine. For example, items produced by the existing machines are brought to the new machine for further processing. Let us assume that the same number of items will be brought to the new machine from each of the four existing machines and that there is a fixed cost per unit distance of such movement for each item. We wish to locate the new machine in such a way that the total cost of this operation will be a minimum.

Mathematically, our problem is to locate the new machine in such a way that the sum of the distances to the existing machines is a minimum. If x denotes the location of the new machine, then our problem becomes the following:
Determine the value of x for which

$$C(x) = |x - (-3)| + |x - (-1)| + |x - 2| + |x - 6|$$

is a minimum. In order to sketch a graph of the function $C(x)$, we consider the following five cases:

1) $x \leq -3$. Then

$$C(x) = (-3 - x) + (-1 - x) + (2 - x) + (6 - x) = -4x + 4.$$

2) $-3 \leq x \leq -1$. Then

$$C(x) = (x + 3) + (-1 - x) + (2 - x) + (6 - x) = -2x + 10.$$

3) $-1 \leq x \leq 2$. Then

$$C(x) = (x + 3) + (x + 1) + (2 - x) + (6 - x) = 12.$$

4) $2 \leq x \leq 6$. Then

$$C(x) = (x + 3) + (x + 1) + (x - 2) + (6 - x) = 2x + 8.$$

61

5) $x \geq 6$. Then

$$C(x) = (x + 3) + (x + 1) + (x - 2) + (x - 6) = 4x - 4.$$

The function $C(x)$ may be written more compactly as follows:

$$C(x) = \begin{cases} -4x + 4, & x \leq -3 \\ -2x + 10, & -3 \leq x \leq -1 \\ 12, & -1 \leq x \leq 2 \\ 2x + 8, & 2 \leq x \leq 6 \\ 4x - 4, & 6 \leq x \end{cases}$$

The graph of $C(x)$ is now given in Figure 2. The units on the y-axis have been chosen to yield a practical graph. It follows that the new machine may be located at any point between $x = -1$ and $x = 2$ since the sum of the distances to the existing machines for any such x is equal to 12, the minimum value of $C(x)$.

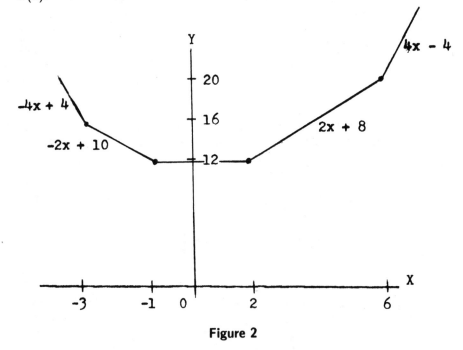

Figure 2

Let us now consider a slightly more complicated example. Again let us assume that four machines in a certain factory are located along an aisle and again there will be interaction between the existing machines and the new machine. For example, items produced by the existing machines are brought to the new machine for further processing. Let us assume that the same number of items will be brought to the new machine from m_2 and m_4. Twice as many items, however, will be brought to the new machine from m_1 as from m_2 and three times as many items will be brought to the new machine from m_3 as from m_2. Again we suppose that there is a fixed cost per unit distance of such movement

for each item. We wish to locate the new machine in such a way that the total cost of this operation will be a minimum.

Since the number of items brought to the new machine varies between the existing machines, we must now assign "weights" to the distances between the new machine and the existing machines. We assign a weight of one to the distance between the new machine and m_2. Since the same number of items will be brought from the machine m_4, we assign a weight of one to the distance between the new machine and m_4. Then the distance between the new machine and m_1 will get a weight of two, while the distance between the new machine and m_3 will get a weight of three. If x denotes the location of the new machine, then our problem becomes the following: Determine the value of x for which

$$C(x) = 2|x - (-3)| + |x - (-1)| + 3|x - 2| + |x - 6|$$

is a minimum. Again we consider five cases:

1) $x \leq -3$. Then

$$C(x) = 2(-3 - x) + (-1 - x) + 3(2 - x) + (6 - x) = -7x + 5.$$

2) $-3 \leq x \leq -1$. Then

$$C(x) = 2(x - (-3)) + (-1 - x) + 3(2 - x) + (6 - x) = -3x + 17.$$

3) $-1 \leq x \leq 2$. Then

$$C(x) = 2(x + 3) + (x + 1) + 3(2 - x) + (6 - x) = -x + 19.$$

4) $2 \leq x \leq 6$. Then

$$C(x) = 2(x + 3) + (x + 1) + 3(x - 2) + (6 - x) = 5x + 7.$$

5) $6 \leq x$. Then

$$C(x) = 2(x + 3) + (x + 1) + 3(x - 2) + (x - 6) = 7x - 5.$$

The function $C(x)$ may be written more compactly as follows:

$$C(x) = \begin{cases} -7x + 5, & x \leq -3 \\ -3x + 17, & -3 \leq x \leq -1 \\ -x + 19, & -1 \leq x \leq 2 \\ 5x + 7, & 2 \leq x \leq 6 \\ 7x - 5, & 6 \leq x \end{cases}$$

The graph of $C(x)$ is now given in Figure 3. It follows that the new machine should be located at $x = 2$. Since an existing machine is already located at $x = 2$, this is physically impossible. However, the results indicate that we should locate the new machine as close to $x = 2$ as is physically possible.

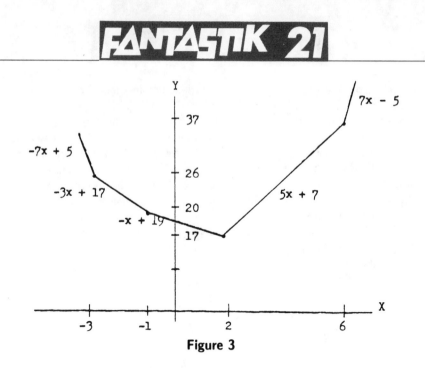

Figure 3

Indeed, we can say more. Since the graph is "steeper" to the right of $x = 2$ than to the left, one should locate the new machine to the left of $x = 2$ and as close to $x = 2$ as is physically possible.

Practice Problem 1:

There are five machines, m_1, m_2, m_3, m_4, and m_5, in a certain factory which are located along an aisle (the x-axis) as illustrated in the following figure:

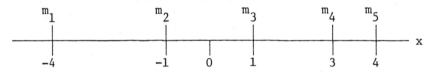

A new machine is to be located somewhere along this aisle. The same number of items will be brought to the new machine from each of the five existing machines and there is a fixed cost per unit distance of such movement for each item. Where should the new machine be located in order to minimize the cost of this operation?

Practice Problem 2:

In the second example of this fantastik, show that the result does not depend on the positions of the existing machines, but only on the weights assigned to each location. In particular, suppose there are four machines m_1, m_2, m_3, and m_4 in a certain factory which are located along an aisle (the x-axis) at $x_1^*, x_2^*, x_3^*, x_4^*$, as illustrated in the following figure:

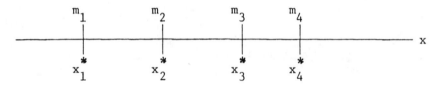

A new machine is to be located along this aisle. The same number of items will be brought to the new machine from m_2 and m_4. Twice as many items, however, will be brought to the new machine from m_1 as from m_2 and three times as many items will be brought to the new machine from m_3 as from m_2. There is a fixed cost per unit distance of movement for each item. Show that the optimal location is as close to m_3 as possible, but to the left of m_3. (*Hint:* Find $C(x)$ as described. Show that $C(x_1^*) > C(x_2^*)$, $C(x_2^*) > C(x_3^*)$, and $C(x_4^*) > C(x_3^*)$.)

A Dear Pier

Mathematics Used: *Arithmetic-Geometric Mean*
Inequality (See Appendix I)

Suppose one is interested in constructing a bridge to cross a river. The span is of total length L ft. A designer has an opportunity to use piers to support the weight but the piers must be equally spaced. There are many possible designs and only several are given in the figures which follow:

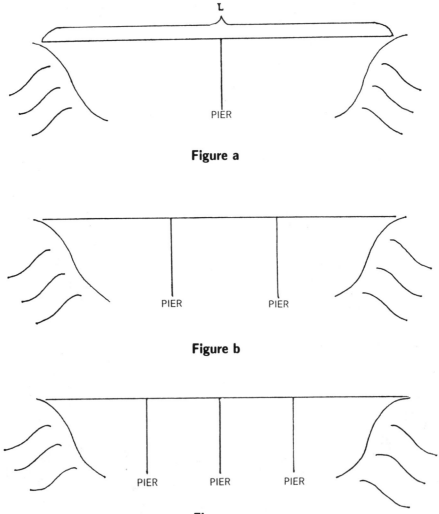

Figure a

Figure b

Figure c

More piers reduce the span length between piers. The smaller span lengths cost less because lighter steel can be used. However, the total cost of making the piers obviously increases as more piers are built.

Let n = number of individual spans. For example, in Figure a the value of n is 2 while in Figure b the value of n is 3. Note that the number of piers is $n - 1$. Also, let

l = length, in feet, of an individual span
P = cost of building one pier
W = pound weight of steel per foot of span
C = cost per pound of bridge steel.

Note that n, l, and L are related by

$$nl = L$$
$$n = \frac{L}{l}.$$

The total cost of making the piers is

$$(n - 1)P = \left(\frac{L}{l} - 1\right)P.$$

As previously mentioned, the weight of steel used depends on the individual span length, say Kl, where $K > 0$ is a constant of proportionality. The cost of steel is thus given by

$$C(Kl)L.$$

The total cost TC of building the bridge is given by

$$TC = \left(\frac{L}{l} - 1\right)P + C(Kl)L$$
$$TC = \frac{L}{l}P - P + C(Kl)L.$$

One wishes to determine l so that TC is as small as possible.

The values of P, L, and K are assumed to be constants. Thus, the total cost T over which one has control is given by

$$T = \frac{L}{l}P + CKlL.$$

A value of l which minimizes T will also minimize TC.

By the arithmetic-geometric mean inequality

$$\frac{\frac{L}{l}P + CKlL}{2} \geq \sqrt{L^2 PKC}$$
$$\frac{T}{2} \geq \sqrt{L^2 PKC}$$
$$T \geq 2\sqrt{L^2 PKC}.$$

The right-hand side is a constant and one can obtain this minimum value of T by having

$$\frac{L}{l}P = CKlL$$

$$l^2 = \frac{P}{CK}.$$

Since $l > 0$, one takes

$$l = \sqrt{\frac{P}{CK}}.$$

It is interesting to note that the value of l which minimizes the total cost does not depend on the total span length L.

   Problem: In the above discussion n is assumed to be a positive integer. The value of n is given by

$$n = \frac{L}{l}.$$

Suppose one computes the value of l which minimizes T (and hence TC) and then computes n. If the resulting value of n is not an integer, how should one handle this situation?

Judge Judges

Mathematics Used: *Algebra, Elementary Probability*

In a certain region of the country, a committee is considering an "optimal" jury size. A decision by a jury is made on the basis of a simple majority. If there are an "even" number of people on a jury, the result could be a tie (hung jury) in which case there is a retrial. Retrials are expensive. Moreover, an incorrect decision on the part of a jury is considered to be inhuman. Thus, "optimal jury size" to this committee means that size which has the greatest chance of making a correct decision on the first trial when compared to other sizes under consideration.

In this section, we shall restrict our considerations to juries of size 2, 3, and 4. Let us suppose that there is a probability p, $0 < p < 1$, of an **individual** making a correct decision. Now, assuming that decisions are made on an independent basis, the probability that a two-person jury will make a correct decision on the first trial is given by p^2.

In the case of a three-person jury, a correct decision results from one of the following:

Person 1	Person 2	Person 3	Probability
right	right	right	p^3
wrong	right	right	$(1-p)p^2$
right	wrong	right	$p(1-p)p$
right	right	wrong	$p^2(1-p)$

Thus, the probability that a three-person jury will make a correct decision on the first trial is given by

$$p^3 + 3p^2(1-p).$$

In the case of a four-person jury, a correct decision on the first trial results from one of the following:

Person 1	Person 2	Person 3	Person 4	Probability
right	right	right	right	p^4
wrong	right	right	right	$(1-p)p^3$
right	wrong	right	right	$p(1-p)p^2$
right	right	wrong	right	$p^2(1-p)p$
right	right	right	wrong	$p^3(1-p)$

Thus, the probability that a four-person jury will make a correct decision on the first trial is given by

$$p^4 + 4p^3(1-p).$$

The information gained thus far is summarized in the following table:

Size of Jury	Probability of a correct decision on the first trial
2	p^2
3	$p^3 + 3p^2(1-p)$
4	$p^4 + 4p^3(1-p)$

Now, under our definition of "optimal jury size", we would use a two-person jury over a three-person jury if

$$p^2 > p^3 + 3p^2(1-p)$$

or (dividing by p^2)

$$1 > p + 3(1-p)$$
$$1 > -2p + 3$$
$$2p > 2$$
$$p > 1. \qquad (1)$$

Since the inequality in (1) is **never** satisfied, it follows that we would never pick a two-person jury over a three-person jury.

We would use a four-person jury over a three-person jury if

$$p^4 + 4p^3(1-p) > p^3 + 3p^2(1-p)$$

or (dividing by p^2)

$$p^2 + 4p(1-p) > p + 3(1-p)$$
$$-3p^2 + 6p - 3 > 0$$
$$p^2 - 2p + 1 < 0$$
$$(p-1)^2 < 0. \qquad (2)$$

Since the inequality in (2) is never satisfied, it follows that we would never pick a four-person jury over a three-person jury.

The above analysis indicates that a three-person jury is "optimal" if one considers two, three, and four-person juries.

Let us conclude this discussion by comparing a two-person jury to a four-person jury. Now, we would use a two-person jury over a four-person jury if

$$p^2 > p^4 + 4p^3(1-p)$$

or (dividing by p^2)

$$1 > p^2 + 4p(1-p)$$
$$3p^2 - 4p + 1 > 0$$
$$(3p-1)(p-1) > 0.$$

We have two cases to consider.

Case I: $3p - 1 > 0$ and $p - 1 > 0$. But the second inequality implies that $p > 1$ which is impossible, and, hence, we can dismiss this case.

Case II: $3p - 1 < 0$ and $p - 1 < 0$. These inequalities imply $3p < 1$ and $p < 1$ or simply $p < \frac{1}{3}$.

We can thus conclude that one would pick a two-person jury over a four-person jury if $p < \frac{1}{3}$ and a four-person jury over a two-person jury if $p > \frac{1}{3}$. If $p = \frac{1}{3}$, then both have the same chance of making a correct decision on the first trial. If $p > \frac{1}{3}$, and in an educated society one would hope that would happen, a four-person jury would be preferable to a two-person jury.

There is one interesting observation concerning the above analysis. We concluded that we would pick a two-person jury over a four-person jury if $p < \frac{1}{3}$. That is, if the probability of an individual making a correct decision is "small", then a two-person jury is preferable to a four-person jury. An equivalent statement is the following: **If the probability of an individual making a wrong decision is "high", then a two-person jury is preferable to a four-person jury**.

Practice Problem 1: Using the definition of "optimal jury size" given in this section, which is preferable, a one-person jury or a three-person jury?

Practice Problem 2: For a certain trip an individual has a choice of a two-engine plane or a four-engine plane. If either engine on the two-engine plane fails, the plane is unable to fly. The four-engine plane, however, is able to fly with only three engines operating, but no less. Which plane would you take if

i) the "trouble rate" of an engine is the same for both planes.

ii) the probability that a given engine will "fail" on this trip is 1/10 for the two-engine plane and 1/12 for the four-engine plane.

Practice Problem 3: Using the definition of "optimal jury size" given in this section, which is preferable, a five-person jury or a three-person jury?

Slice the Price

Mathematics Used: *Quadratic Functions*

In algebra texts, one frequently encounters problems such as the following:

A manufacturer has determined that if his selling price for a certain
item is x (dollars), then his yearly profit P on such items is given by

$$P = -20x^2 + 7000x - 300,000.$$

Determine the value of x for which the profit is a maximum.

Such problems come off as "contrived" to many students. It is possible,
however, to make such problems more realistic.

When a company has to make a decision on whether or not to produce a
new product, marketing experts are normally consulted as to the demand for
this product. Demand for a product is frequently related to the selling price in
the following way: As the selling price increases, the demand decreases. Let us
suppose that the marketing experts have presented us with the following data:

Selling Price (in dollars)	Yearly Demand
s	d
50	5000
100	4000
150	3000
300	0

Let us suppose that it costs $50 to produce each item so that the company is
certainly not willing to consider any selling price below $50. If we plot the above
data, we obtain the following:

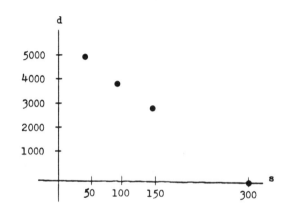

It is not difficult to show that these "data points" lie on a straight line and, using any two of the points, the equation of the straight line is given by

$$d = -20s + 6000. \tag{1}$$

The relevant segment is graphed below.

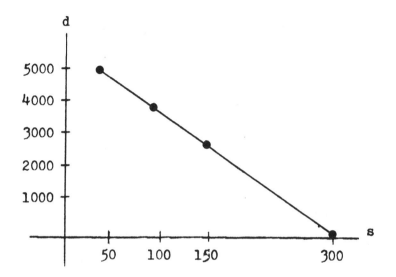

In general, one cannot expect the data points to lie on a straight line, but it is frequently the case that the data points "come close" to lying on a straight line. In this case, one can determine an equation such as (1) by finding the straight line which is the "best fit" in the sense of least squares. (See figure below.)

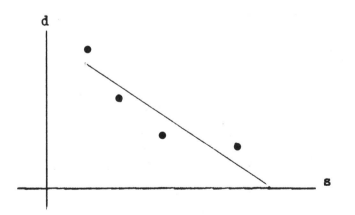

Now, letting P denote the yearly profit, we have

$$P = \text{Total Revenue} - \text{Production Costs}$$
$$P = sd - 50d$$
$$P = (s - 50)d.$$

Using (1), we have

$$P = (s - 50)(-20s + 6000)$$
$$P = -20s^2 + 7000s - 300,000. \qquad (2)$$

(Compare this result with the problem given at the beginning of this section.)

Our problem is now to determine a selling price $s, s \geq 50$, which will maximize the yearly profit P. One way of solving this problem is the following:

$$P = -20s^2 + 7000s - 300,000$$
$$= -20s[s^2 - 350s] - 300,000$$

(completing the square)

$$= -20[s^2 - 350s + (175)^2] + 312,500$$
$$P = -20[s - 175]^2 + 312,500. \qquad (3)$$

From (3), it is clear that we should take $s = 175$ if we wish to maximize our yearly profit. A graph of s versus P is given below.

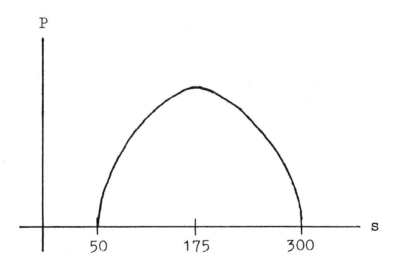

Practice Problem 1: For a new item, the marketing experts have estimated the following:

Selling Price (in dollars) s	Yearly Demand d
100	7000
300	5000
600	2000
800	0

The production cost per item is $40. Determine a selling price which optimizes the yearly profit.

Practice Problem 2: One should observe that in the example of this section, as well as in practice problem 1, the optimal selling price is simply the average of the minimum and maximum selling prices. Show that under the assumptions of the example, this will always be the case. (Let $d = As + B$ where $A < 0$ and $B > 0$ and let C be the cost of producing one item.)

Practice Problem 3: Suppose that marketing experts have presented you with the following data:

Selling Price (in dollars)	Yearly Demand
s	d
50	5000
100	35000
150	3000
300	0

Also, suppose that it costs $50 to produce each item so that the company is not willing to consider any selling price below $50. How would you determine a selling price in order to optimize the yearly profit? What is that price?

The Bore War

Mathematics Used: *Algebra, Elementary Probability*
(Review Fantastik 18)

In order to get some insight into how mathematics is used in making decisions on war strategy, we consider the following "contrived" war between two small countries A and B. We assume the following:

1. Country A has two planes, and there are two air routes from A to B. In country B, there is a small bridge which is vital to B's military efforts. The two planes of country A are to be used to destroy the bridge. The bridge requires about 24 hours to rebuild and each plane makes one daily flight in an attempt to keep the bridge in an unusable condition. If a plane is shot down, a large "neutral" power will immediately supply country A with a new plane.

2. Country B has two anti-aircraft guns which it uses along the air routes in an attempt to shoot down the planes of country A.

3. As there are two routes from A to B, country A can send both planes along one route or one plane along each route.

4. Country B can place both guns along one route or place one gun along each route.

5. If one plane (two planes) travel(s) along a route on which there is one gun (two guns), then that plane (both planes) will be shot down. However, if two planes travel along a route on which there is one gun, then only one plane will be shot down.

6. If a plane gets through to the target, then the target will be destroyed.

The problem, of course, is to determine, in some sense, "optimal" strategies for countries A and B. Let D denote the strategy "use different routes" and S denote the strategy "use same route".

Now, on a given day, if both countries use strategy D, then there is 0 "payoff" to country A, since neither plane will reach the target. If country A uses D and country B uses S, then at least one plane will get through, and the probability of the target being destroyed is 1. If A uses S and B uses D, then, again, one plane will get through and the probability of the target being destroyed is 1. If both countries use S, then there is a $\frac{1}{2}$ chance that country A will pick the route with guns and, hence, there is a probability of $\frac{1}{2}$ that the target will be destroyed.

We now put this data into the standard form of a game.

$$
\begin{array}{cc}
 & B \\
 & \begin{array}{cc} (q) & (1-q) \\ D & S \end{array}
\end{array}
$$

$$
A \quad
\begin{array}{c}
(p) \quad D \\
(1-p) \quad S
\end{array}
\begin{pmatrix}
0 & 1 \\
1 & 1/2
\end{pmatrix}
$$

The expected pay-off to A, denoted E, is given by

$$E = (0)pq + (1)p(1-q) + (1)(1-p)q + \frac{1}{2}(1-p)(1-q)$$

$$E = -\frac{3}{2}pq + \frac{1}{2}p + \frac{1}{2}q + \frac{1}{2}$$

$$E = -\frac{3}{2}\left[pq - \frac{1}{3}p - \frac{1}{3}q\right] + \frac{1}{2}$$

$$E = -\frac{3}{2}\left(p - \frac{1}{3}\right)\left(q - \frac{1}{3}\right) + \frac{2}{3}.$$

From our analysis of such games, we see that an "optimal" strategy for A is to choose $p = \frac{1}{3}$. That is, A should send the two planes along different routes $\frac{1}{3}$ of the time and along the same route $\frac{2}{3}$ of the time. This results in a "pay-off" of $\frac{2}{3}$ to A which indicates that country A can expect the target to be destroyed about $\frac{2}{3}$ of the time. A similar analysis shows that country B should choose $q = \frac{1}{3}$.

A New Pair of Genes

Mathematics Used: *Algebra, Elementary Probability*

Biological information concerning hereditary characteristics is transmitted by *genes* (the study of genetics). We consider only two gene forms, A and a. An individual receives one of these forms from each parent. Thus, an individual can be classified as AA, or aA (or Aa), or aa.

The following work is due to Hardy-Weinberg. We assume that

1) for each classification, A, or aA, or aa, there are an equal number of males and females.
2) All members of the population are equally fertile. In particular, an individual of type AA is just as likely to reproduce as an individual of type aA, etc.
3) Mating is random among classifications.
4) If an individual is of type aA, then the probability of transmitting an a (or A) during mating is one-half.

Suppose at a certain time the proportions of AA, aA (or Aa), and aa individuals are given by p_0, q_0, and r_0, respectively. For example, if $p_0 = 4/10$, then this means that 40% of the population is of type AA. One may also interpret p_0 as the probability that an individual selected at random is of type AA. In particular, we must have $p_0 + q_0 + r_0 = 1$.

We now wish to calculate the proportions in the next generation. The proportions in this generation shall be designated as p_1, q_1, and r_1. We shall refer to this generation as the "first" generation.

In order to obtain an AA type in the first generation, we must have one of the following:

i) an AA male mating with an AA female. This yields a probability of p_0^2.
ii) an AA male mating with an aA female, the female donating A, or an aA male mating with an AA female, the male donating A. This yields a probability of $p_0 q_0$.
iii) an aA male mating with an aA female, each donating an A. This yields a probability of $\frac{1}{4} q_0^2$.

The following table can be a conceptual aid.

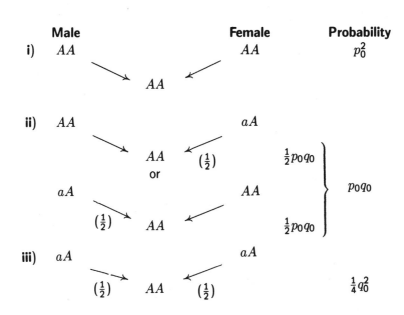

	Male		Female	Probability

Hence,

$$p_1 = p_0^2 + p_0 q_0 + \frac{1}{4} q_0^2$$

$$p_1 = \left(p_0 + \frac{1}{2} q_0 \right)^2.$$

In order to obtain an aA type in the first generation, we must have one of the following:

i) an AA male mating with an aa female. This yields a probability of $2p_0 r_0$.

ii) an AA male mating with an aA female, the female donating an a, or an aA male mating with an AA female, the male donating an a. This yields a probability of $p_0 q_0$.

iii) an aa male mating with an aA female, the female donating an A, or an aA male mating with an aa female, the male donating an A. This yields a probability of $q_0 r_0$.

iv) an aA male mating with an aA female, the male donating a and the female A, or the male donating A and the female a. This yields a probability of $\frac{1}{2} q_0^2$.

Male		Female	Probability

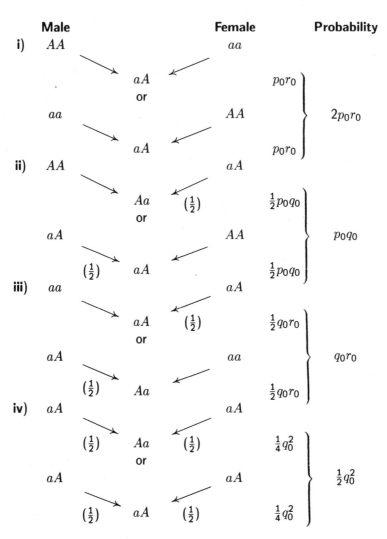

i) AA aa

aA $p_0 r_0$

or

aa AA $2p_0 r_0$

aA $p_0 r_0$

ii) AA aA

Aa $\left(\frac{1}{2}\right)$ $\frac{1}{2}p_0 q_0$

or

aA AA $p_0 q_0$

$\left(\frac{1}{2}\right)$ aA $\frac{1}{2}p_0 q_0$

iii) aa aA

aA $\left(\frac{1}{2}\right)$ $\frac{1}{2}q_0 r_0$

or

aA aa $q_0 r_0$

$\left(\frac{1}{2}\right)$ Aa $\frac{1}{2}q_0 r_0$

iv) aA aA

$\left(\frac{1}{2}\right)$ Aa $\left(\frac{1}{2}\right)$ $\frac{1}{4}q_0^2$

or

aA aA $\frac{1}{2}q_0^2$

$\left(\frac{1}{2}\right)$ aA $\left(\frac{1}{2}\right)$ $\frac{1}{4}q_0^2$

Hence,

$$q_1 = 2p_0 r_0 + p_0 q_0 + q_0 r_0 + \frac{1}{2}q_0^2$$

$$q_1 = 2\left(p_0 + \frac{1}{2}q_0\right)\left(r_0 + \frac{1}{2}q_0\right).$$

In order to obtain an aa type in the first generation, we must have one of the following:

i) an aa male mating with an aa female. This yields a probability of r_0^2.

ii) an aA male mating with an aa female, the male donating an a, or an aa male mating with an aA female, the female donating an a. This yields a probability of $q_0 r_0$.

iii) an aA male mating with an aA female, each donating an a. This yields a probability of $\frac{1}{4}q_0^2$.

The following table can be conceptual aid.

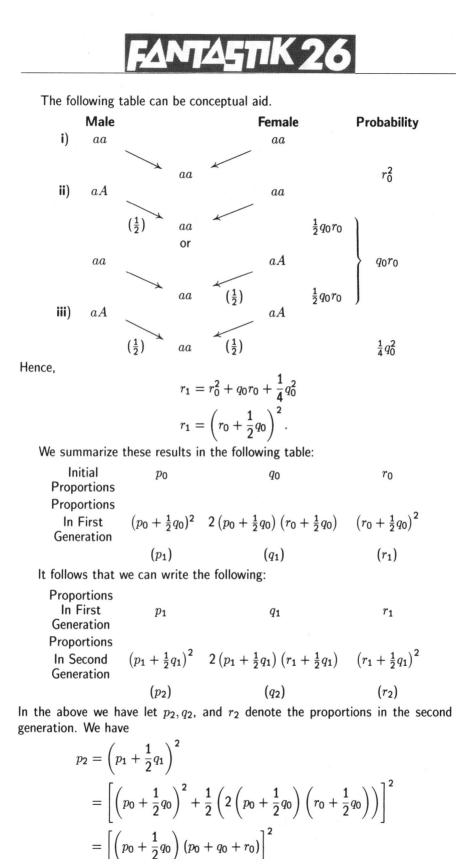

Male		Female	Probability

i) aa and aa → aa, probability r_0^2

ii) aA $(\frac{1}{2})$ → aa ← aa, $\frac{1}{2}q_0 r_0$

or

aa → aa ← $(\frac{1}{2})$ aA, $\frac{1}{2}q_0 r_0$ ⎫ $q_0 r_0$ ⎭

iii) aA $(\frac{1}{2})$ → aa ← $(\frac{1}{2})$ aA, $\frac{1}{4}q_0^2$

Hence,

$$r_1 = r_0^2 + q_0 r_0 + \frac{1}{4}q_0^2$$

$$r_1 = \left(r_0 + \frac{1}{2}q_0\right)^2.$$

We summarize these results in the following table:

	p_0	q_0	r_0
Initial Proportions	p_0	q_0	r_0
Proportions In First Generation	$(p_0 + \frac{1}{2}q_0)^2$	$2\left(p_0 + \frac{1}{2}q_0\right)\left(r_0 + \frac{1}{2}q_0\right)$	$\left(r_0 + \frac{1}{2}q_0\right)^2$
	(p_1)	(q_1)	(r_1)

It follows that we can write the following:

	p_1	q_1	r_1
Proportions In First Generation	p_1	q_1	r_1
Proportions In Second Generation	$\left(p_1 + \frac{1}{2}q_1\right)^2$	$2\left(p_1 + \frac{1}{2}q_1\right)\left(r_1 + \frac{1}{2}q_1\right)$	$\left(r_1 + \frac{1}{2}q_1\right)^2$
	(p_2)	(q_2)	(r_2)

In the above we have let p_2, q_2, and r_2 denote the proportions in the second generation. We have

$$p_2 = \left(p_1 + \frac{1}{2}q_1\right)^2$$

$$= \left[\left(p_0 + \frac{1}{2}q_0\right)^2 + \frac{1}{2}\left(2\left(p_0 + \frac{1}{2}q_0\right)\left(r_0 + \frac{1}{2}q_0\right)\right)\right]^2$$

$$= \left[\left(p_0 + \frac{1}{2}q_0\right)(p_0 + q_0 + r_0)\right]^2$$

or (since $p_0 + q_0 + r_0 = 1$)

$$p_2 = \left(p_0 + \frac{1}{2}q_0\right)^2.$$

Observe that $p_2 = p_1$.

Similarly,

$$q_2 = 2\left(p_1 + \frac{1}{2}q_1\right)\left(r_1 + \frac{1}{2}q_1\right)$$

$$= 2\left[\left(p_0 + \frac{1}{2}q_0\right)^2 + \frac{1}{2}\left(2\left(p_0 + \frac{1}{2}q_0\right)\left(r_0 + \frac{1}{2}q_0\right)\right)\right]$$

$$\cdot\left[\left(r_0 + \frac{1}{2}q_0\right)^2 + \frac{1}{2}\left(2\left(p_0 + \frac{1}{2}q_0\right)\left(r_0 + \frac{1}{2}q_0\right)\right)\right]$$

$$= 2\left[\left(p_0 + \frac{1}{2}q_0\right)(p_0 + q_0 + r_0)\right]\left[\left(r_0 + \frac{1}{2}q_0\right)(p_0 + r_0 + q_0)\right]$$

or (since $p_0 + q_0 + r_0 = 1$)

$$q_2 = 2\left(p_0 + \frac{1}{2}q_0\right)\left(r_0 + \frac{1}{2}q_0\right).$$

Observe that $q_2 = q_1$.

Similarly,

$$r_2 = \left(r_1 + \frac{1}{2}q_0\right)^2$$

$$= \left[\left(r_0 + \frac{1}{2}q_0\right)^2 + \frac{1}{2}\left(2\left(p_0 + \frac{1}{2}q_0\right)\left(r_0 + \frac{1}{2}q_0\right)\right)\right]^2$$

$$= \left[\left(r_0 + \frac{1}{2}q_0\right)(p_0 + q_0 + r_2)\right]^2$$

or (since $p_0 + q_0 + r_0 = 1$)

$$r_2 = \left(r_0 + \frac{1}{2}q_0\right)^2.$$

Observe that $r_2 = r_1$.

That is, the proportions of AA, aA, and aa have reached "equilibrium" by the second generation.

Practice Problem: Suppose at a certain time, the proportions of AA, aA (or Aa), and aa are given by .3, .6, and .1, respectively. What will the proportions be in the next generation?

Cues on Queues

Mathematics Used: *Algebra, Inequalities*

The theory of queues (waiting lines) now represents a relatively large area of mathematics. One can observe queues in a shopping market, at a restaurant, or at a toll booth on a turnpike. In general, customers do not arrive at such "service facilities" at a constant rate. Thus, the theory of probability very quickly dominates the mathematics involved. However, there are some situations where customers (not necessarily human beings) do arrive at a service facility at a constant rate. For example, the author recently toured a metal-craft company and observed a machine which was used to "smooth" each raw item. These items were then taken, via a belt, to another machine where a lacquer finish was sprayed on each item. One could regard the smooth items as "customers" for the spray machine and the spray machine as a service facility. In this case, customers arrived at a constant rate. Now, suppose the spray machine breaks down. Then a queue is going to form while the machine is being repaired. A problem that arises is the following: Given that we have some idea of how long it will take to repair the machine, about how long will it take to eliminate the resulting queue once the machine again becomes operational?

One can gain some insight into the problem by considering items which come from a machine A to a machine B at the rate of 30 items per hour. Machine B is *capable* of servicing 80 items per hour. Suppose machine B breaks down and is inoperative for four hours. When the machine begins to operate again, a waiting line of 120 items has developed. Now, during the first hour (after machine B again begins operation) 30 new items arrive at B, but machine B has been able to service 80 items. Thus, at the end of one hour, there is a waiting line of 70 items. During the second hour, 30 new items arrive at B, but B can service 80 items. Thus, at the end of two hours, there is a waiting line of 20 items. During the third hour, 30 new items arrive at B, but B can service 80 items. Thus, during the third hour, the waiting line will disappear and production will return to normal (see table below).

Time Factor	Waiting Line
Machine B begins operation	120
End of one hour	70
End of two hours	20
End of three hours	0

(waiting line disappears during third hour)

Let us now consider the following situation. We have a service facility which requires two minutes to service a customer. A new customer arrives every five minutes. When the service facility begins operation there are six customers waiting and the first "new" customer arrives one minute later. Let $e_1, e_2, e_3, e_4, e_5, e_6$ denote the customers in the waiting line when the operation begins ("e" for early bird) and let c_1, c_2, c_3, etc. denote the "new" customers that arrive, in the order indicated by the subscripts. The following line indicates arrival and departure times.

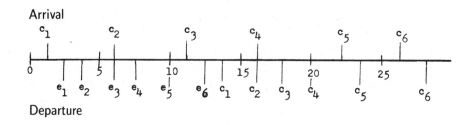

We see that c_5 is the first customer to arrive with no queue present. Moreover, 20 minutes were required to dispel the queue. Here the queue consists of the customers in the waiting line **or** in the service facility.

One can imagine that for most such problems, an arrival-departure line, such as that used above, is quite impractical. Therefore, we seek another method for obtaining the desired information.

For the process described above, let c_{n+1} be the first customer to arrive with no queue present. This means that the service facility has handled $n + 6$ customers by this time. The total time in minutes needed to serve these $n + 6$ customers is

$$(n + 6) \cdot 2.$$

However, when customer $(n + 1)$ arrives, the facility has been operating for

$$n \cdot 5 + 1$$

minutes. Hence, we must have

$$(n + 6) \cdot 2 \leq n \cdot 5 + 1$$
$$11 \leq 3n$$
$$\frac{11}{3} \leq n. \tag{1}$$

Note that n must be an integer. In addition, we want the **first** customer to arrive and find no queue. Thus, we want the smallest integer n which satisfies (1). It follows that $n = 4$ and, hence, the fifth customer c_5 is the first customer to arrive and find no queue. The total time required to dispel the queue is simply the time required to service the $6 + 4 = 10$ previous customers which, of course, is 20 minutes.

We now generalize the above process by letting N denote the number of people in the waiting line when the service facility opens. Let S denote the time required to service a customer and let T denote the time between arrivals. Let f denote the time between the opening of the service facility and the arrival of the first new customer, $0 \leq f < T$.

Observe that for the queue to disappear we must have $T > S$. For if $T < S$, then the queue will simply continue to grow. For example, observe what happens

if a machine processes 100 items per hour and sends them for further processing to a machine that can process only 60 items per hour. If $T = S$, the queue remains standing and if $N \neq 0$, the queue will never disappear. **Hence, we shall assume that $T > S$.**

Now, let c_{n+1} be the first customer to arrive with no queue present. This means that the service facility has handled $n + N$ customers. The total time needed to serve these $n + N$ customers is

$$(n + N)S.$$

However, when customer $(n + 1)$ arrives, the time that the facility has been operating is given by

$$nT + f.$$

Hence, we must have

$$(n + N)S \leq nT + f$$
$$NS - f \leq n(T - S)$$
$$\frac{NS - f}{T - S} \leq n. \tag{2}$$

(Note that we use here the fact that $T > S$ or $T - S > 0$.) Thus, n is the smallest integer which satisfies (2). The first customer to arrive with no queue present is c_{n+1} and the total time required for the queue to disappear is given by

$$(n + N)S.$$

Practice Problem 1: Suppose a single queue is serviced by two facilities each of which requires a time of S to service a customer. Using the notation in our generalization above, determine the customer will be the first to arrive with no queue present and the total time required for the queue to disappear.

Practice Problem 2: In the generalization of this section, we assumed that $0 \leq f < T$. What happens if $f \geq T$?

Practice Problem 3: In the generalization of this section, determine how long a given customer will wait for completion of service.

A Bloody Affair

Mathematics Used: *Algebra (Review Fantastiks 8 and 9)*

In Fantastiks 8 and 9 we introduced the idea of a linear programming problem. In this section, we shall show how one can take a fairly difficult problem and "set it up" as a linear programming problem. We shall not solve the problem, but shall assume that a computer program is available to do the job, if necessary.

Let us assume that a hospital has available the following:

Blood Type	Supply (Pints)	Cost/Pint of Replacement
A	8	$25
B	5	$25
AB	2	$40
O	6	$20

There are six patients with the following needs for transfusions.

Patient	Blood Type	Pints Required
1	A	4
2	B	2
3	B	1
4	AB	3
5	O	1
6	O	3

The problem is to meet the patients' needs while minimizing replacement costs.

Now, an A-type donor can give blood to an A or AB-type recipient. A B-type donor can give blood to a B or AB-type recipient. An AB-type donor can only give blood to an AB-type recipient. An O-type donor is a "universal" donor in the sense that an O-type donor can give blood to a recipient of type A, B, AB, or O.

Suppose we form the following schematic representation:

Recipient

Donor	1	2	3	4	5	6	less than or equal to
A	a			g			8
B		c	e	h			5
AB				j			2
O	b	d	f	k	l	m	6
greater than or equal to	4	2	1	3	1	3	

where $a, b, c, \ldots, m$ are the number of pints going from donors to the noted recipients. Thus, the boundary inequalities relative to this problem are

$$\begin{cases} a + g \leq 8 \\ c + e + h \leq 5 \\ j \leq 2 \\ b + d + f + k + m \leq 6 \\ a + b \geq 4 \\ c + d \geq 2 \\ e + f \geq 1 \\ g + h + j + k \geq 3 \\ l \geq 1 \\ m \geq 3 \\ a \geq 0, \ b \geq 0, \ldots, m \geq 0 \end{cases}$$

and we note that this describes a geometric figure in 12-dimensional space.

The cost of replacement is given by

$$C = 25(a + g) + 25(c + e + h) + 40(j) + 20(b + d + f + k + l + m).$$

Thus, for each ordered 12-tuplet $(a, b, c, \ldots, m)$ which satisfies the inequalities above, we compute the value of C, and from all these choices we wish to find that 12-tuplet which makes C the smallest. This is the optimal choice. Problems of this type (which includes the problems of sections 8 and 9) are known as "linear programming" problems.

Because of the inequalities, the figure described is a "finite" figure of 12-dimensional space, as are those figures in 2-space which can be enclosed in a square, and analogously those in 3-spaces which are in a cube. For such figures, a function like C assumes a minimum value and that value occurs at a "corner". Thus, one need only find the values at all of the corners. There are, for the problem under discussion, approximately 60 such corners. The process of finding all these corners, let alone finding the value of C, is laborious. The mathematician George Dantzig found an algorithmic process for finding the solution and a process which generally reduced the number of steps required. The technique involves beginning at any corner and moving from that corner to an "adjacent" corner which makes C smaller. By such a process, one finds the minimum value.

Practice Problem: Given the following data, set up a linear programming problem for minimizing replacement cost.

Blood Type	Supply (Pints)	Cost/Pint of Replacement
A	9	$20
B	6	$20
AB	2	$35
O	7	$15

Patient	Blood Type	Pints Required
1	A	3
2	B	3
3	AB	2
4	O	1

Cite a Satellite

Mathematics Used: *Trigonometry*

Suppose that a satellite is to be put in an equatorial orbit 300 miles above the earth. It is desired to locate tracking stations along the equator. Each tracking station has a scanning screen which covers 180° with the horizon as illustrated in the figure below (Figure 1).

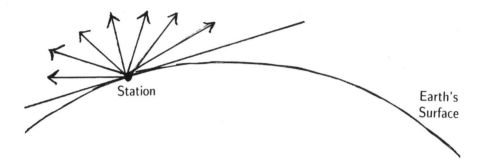

Figure 1

The problem is to locate tracking stations close enough to each other so that we do not have "blind spots" where the satellite is not being observed by at least one scanner, as in Figure 2.

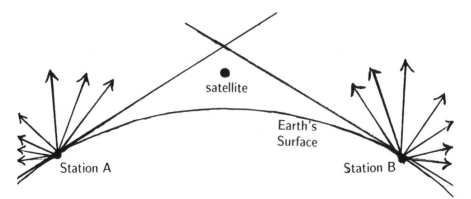

Figure 2

Now, the furthest apart that stations A and B could be is illustrated in Figure 3.

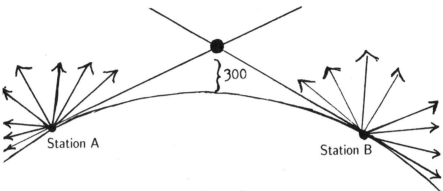

Figure 3

Using 4000 miles as the radius of the earth and observing the symmetry in Figure 3, we see that it is necessary to find the distance d in Figure 4.

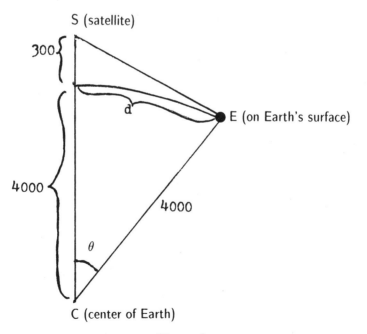

Figure 4

We know that $\angle CES$ is a right angle and, hence,

$$\cos\theta = \frac{4000}{4300} = .9302.$$

Thus,

$$\theta = 21.53°$$
$$\theta = .376 \text{ radians.}$$

It follows that

$$\frac{d}{4000} = .376$$
$$d = 1504 \text{ miles.}$$

It follows that the distance between the two tracking stations cannot be greater than 3008 miles.

Practice Problem: A satellite is in an equatorial orbit 300 miles above the earth and is supposed to orbit the earth once every two hours. Two tracking stations, say A and B, are located 3000 miles apart on the equator. Tracking station B first encounters the satellite at 12 noon with a beam aimed at 30° with the horizon. Tracking station A has a beam aimed at 40° with the horizon (see figure below). If the satellite is in its proper orbit, at what time should the beam from station A intersect the satellite?

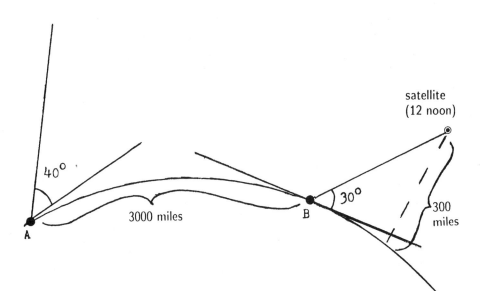

A Test Quest

Mathematics Used: *Elementary Probability, Binomial Tables*

Suppose one is interested in constructing a test with ten questions. The test is to be of the "multiple choice" type with precisely one of the given answers being correct. A student will pass the exam if six or more circled answers are correct, and fail otherwise. The person constructing the exam wants the following: The chance that a student will pass by simply guessing is less than 5/100. The problem is to determine how many answers should be offered to each question so that this condition is satisfied.

Now, suppose that there are m answers for each question. Then the probability that a student will get the correct answer to one question by simply guessing is $1/m$. Using the binomial distribution, the probability that such a student will pass this exam is given by

$$\sum_{k=6}^{10} \binom{10}{k} \left(\frac{1}{m}\right)^k \left(1 - \frac{1}{m}\right)^{10-k} .$$

Hence, we wish to determine m so that

$$\sum_{k=6}^{10} \binom{10}{k} \left(\frac{1}{m}\right)^k \left(1 - \frac{1}{m}\right)^{10-k} < .05.$$

From binomial tables (e.g., *CRC Standard Mathematical Tables*, Ed. Samuel M. Selby, Chemical Rubber Company, Cleveland, Ohio), one finds that

$$\sum_{k=6}^{10} \binom{10}{k} (.3)^k (.7)^{10-k} = .0473,$$

while

$$\sum_{k=6}^{10} \binom{10}{k} (.35)^k (.65)^{10-k} = .0949.$$

It follows that our criteria will be satisfied if

$$\frac{1}{m} \leq .3$$
$$\frac{10}{3} \leq m.$$

Since m must be an integer, we take $m = $ **4**. That is, four answers to each question will be sufficient to guarantee that the probability of a student passing this exam by simply guessing is less than 5/100.

Practice Problem: On a ten-question "true-false" type exam, how should a passing performance be determined so that the chance of a student passing this exam by simply guessing is less than 5/100?

91

Tales of Scales

Mathematics Used: *Ratios*

Author: Howard Hand
         Choate Rosemary Hall
         Wallingford, Connecticut

Background

One cannot speak of music without understanding some fundamentals of the production and transmission of sound, but for present purposes this discussion will be kept to a minimum. Sound is the result of the vibration of some medium, be it wood or metal or another, and the transmission of these vibrations through the air in the form of wavelike increases and decreases in the pressure of the air. These pressure differentials are detected by the eardrum, transformed into neural impulses and thus carried to the brain. The important feature of this transmission for this analysis of music is the **frequency** of the vibration (and thus the frequency of the transmitting wave). These frequencies are commonly measured in cycles per second which are called *hertz*. Whether or not a sound is audible to the normal human ear depends on its frequency, its intensity, and on the relation between the two. Music, however, uses a spectrum roughly between 30 and 4200 hertz.

Sounds of a single frequency are uncommon. For the most part we hear sounds which represent a jumble of frequencies. The tuning fork is a device meant to produce sounds of only one frequency, and such a sound could be called a *pure* musical tone. There are in principle, then, over 4000 tones from which to choose to compose musical **scales**, that is, the particular frequencies of sound chosen to make up melodies and harmonies. In practice, of course, scales are not made up of 4000 or even 1000 musical tones. The piano, for example, one of the most flexible of musical intruments,[1] has just eighty-eight keys. Of these eighty-eight, only twelve have musically different names. The others stand in a relation called the octave (or octaves) to these basic twelve and are a repetition of them. Two notes are in the relation of an octave if the frequency of one is the frequency of the other multiplied by two. More will be said about this later. Why out of all these thousands of possibilities, did there come to be only twelve different notes? This is the question which this article will address.

Harmonics and Homophony

Again, one rarely hears simple sounds, those consisting of vibrations of a single frequency. This lack of simplicity is not only a matter of choice, for even when one chooses one's instrument carefully—let us say one decides to pluck a carefully tuned string in the hope that it will vibrate simply—one finds that nature will not cooperate. For the string, left to vibrate freely, will produce

---

[1]Another constraint, of course, is the means by which the music will be produced. The human voice is a popular choice of means, and the effective singing range for voices ranging from bass to soprano is in the neighborhood of 80 to 900 hertz.

several audible frequencies of vibration as well as others which are too faint to be audible. These extra frequencies, called overtones or *harmonics*, stand in a particular relation to the fundamental tone—namely, their frequencies are all integral multiples of the frequency of this fundamental one. Thus, if a string is tuned to vibrate at 440 hertz, the modern international standard for the A above middle C, then there will also be vibrations produced at 880, 1760, 3520,... hertz. The relative audibility of these harmonics depends on the vibrating medium so that harmonics, although present in any medium, will be audible in various proportions. The variation in these proportions gives the distinctive quality to the sound of a musical instrument called its *timbre*.[2]

As long as one is content to produce music which consists of a succession of simple tones (remembering, of course, that tones will never be truly simple), then the choice of frequencies for one's scale is less complicated. Such music, wherein no two notes will ever be played simultaneously, is called *homophonic*. In such music, one is not much concerned with the consonance of tones, that is, how two tones will sound together. There are, however, rules in the traditional harmony of western art music about pleasant and unpleasant successions of tones. Small jumps from one note to another are preferred to large ones, for example. But consider that a jump from one frequency to another that is too small, a change of frequency of only one or two hertz, might not be easily detected. One's choice of scale does, then, have a number of constraints: range of audible frequency, range of frequency which it is practical to produce on the desired instrument, the need to have tones which are neither too close together nor too far apart, and finally, perhaps, the need not to have too many different tones, since the complexity of the instrument and of performance increase with the number of tones to be played.

Harmony

Most of the music which is of interest to modern listeners is not homophonic, but rather it is called *polyphonic*. It consists of two or more notes sounding simultaneously. *Harmony* is precisely the study of those combinations of tones which sound pleasantly together.

In the fifth century B.C., the Greek philosopher and mathematician Pythagoras, while investigating harmony, made some remarkable discoveries. He had, to begin with, understood the first of what are now called Mersenne's laws. This first law states that "when a string and its tension remain unaltered, but the length is varied, the period of vibration is proportional to the length."[3] This means that the **frequency** of the vibration will be inversely proportional to the length of the string. A string of half the length of another will vibrate at twice its frequency. With this principle in mind, Pythagoras began to experiment by dividing a string at constant tension into different ratios of its whole length and then by listening to determine which ratios produced harmonious pairs of tones.

For example, when Pythagoras chose to fix a point on the string which was 2/3 of the way from one end to the other so that the two segments thus produced

---

[2]Sir James Jeans, *Science and Music* (New York, Dover Publications, Inc., 1968), p. 84.
[3]Sir James Jeans, "Mathematics of Music" in *The World of Mathematics*, ed. James R. Newman (New York, Simon and Schuster, 1956), p. 2297.

would be in the ratio of 2:1, he discovered by plucking both parts of the string that the two sounds were remarkably consonant. The interval between these sounds, as we already noted, is now called the octave. He could have been listening, hypothetically, to middle C and also to the C above middle C. (These names for the notes are, of course, modern.) This ratio of frequencies indeed describes what some would consider to be the **most** consonant of intervals. Having this initial success with simple ratios, he might naturally have wondered about the next simplest, the ratio of 3:2. Indeed, two notes produced by strings whose lengths are in the ratio of 3:2 again produce an extremely pleasing pair, that which we now call the interval of a fifth. Continuing to the ratio of 4:3 (since 4:2=2:1), we find a third pleasant and important harmony, the interval of the fourth. In the modern scale, with C as the fundamental, G above C is the fifth and F above C is the fourth. Notice, however, that if we choose to go a fifth above the F, i.e., $(4/3)(3/2) = 2$, we arrive at a tone which is double the frequency of the original C, namely, its octave, so that C is at an interval of a fifth from F. In musical language, the fourth is the *inversion* of the fifth. Going down an interval of a fourth brings us to the same nominal note as going up a fifth. These three tones, C, F, and G, are therefore very importantly linked by the simplicity of the ratios of their frequencies as well as by their euphony. They formed the basis for even the most primitive of harmonies. Why do tones produced by frequencies which stand in **simple** ratios turn out to be so pleasing to the ear? The answer is still a subject for debate.

Having had such success in the discovery of the relationship of these three tones, Pythagoras seems to have decided that he had enough basic information in order to construct his scale based on this single principle, namely, that every note in the scale should have another note, its fifth, also available in the scale. The C and F were already provided for, since G is the fifth above C, and C is the fifth above F. But what about G? Another note should be constructed so that its frequency would be in the relation of 3:2 to the frequency of G.

For convenience, let us imagine that the frequency of the middle C with which we began was 100. (In reality, it is set by common agreement at 261.63 hertz.) Then the G above it, whose frequency is in the ratio of 3:2, will be 150 hertz. Therefore, the fifth above G should have the frequency of $(3/2)(150)$ or $(3/2)^2(100) = 900/4 = 225$. But notice that this new note, which we now call D, is more than an octave above our original C, rather a large leap from our starting point (and we agreed that smaller leaps were preferable). Thus it seems natural to multiply this 225 by 1/2 in order to move the D down an octave and put it closer to our original C. The result is $(9/4)(1/2)(100)=(9/8)(100)$. The new note, D, has a frequency which stands in the ratio of 9:8 to the first note, middle C.

What is next? Invent a note which will be a fifth above D, that is, at a frequency of $(3/2)^2(100)$. Then move it down one octave as we did before, which gives the note (called A) at a frequency ratio of 27/16 to middle C. Continuing this process once more, multiplying the frequency of A by 3/2 and then moving it down an octave, we produce the note E at a frequency of 81/64 or $3^4/2^6$ times the frequency of middle C.

The following table shows the calculation of the frequency ratios of the first six notes in the scale starting with middle C. Remember that F could be produced by going a fifth below C (multiplying by 2/3) and then going up an octave (multiplying by 2) producing the frequency ratio of 4:3.

Table I

Generation of the chromatic scale by the Pythagorean method

Modern name for tone	Ratio of frequency to frequency of C	
C	1	1
G	$(3/2)$	1.5
D	$(3/2)^2(1/2)$	1.125
A	$(3/2)^3(1/2)$	1.6875
E	$(3/2)^4(1/2)^2$	1.2656
B	$(3/2)^5(1/2)^2$	1.8984
F sharp	$(3/2)^6(1/2)^3$	1.423
. . .		

How long shall we continue in this manner? There is nothing to stop us from going on forever except that, as we already noted, it becomes very complicated to perform scales which have too many notes in them. In fact, much very old music is played on scales consisting of only five tones, called a pentatonic scale. These tones stand in relation to each other as do the first five which we produced by the Pythagorean method, namely, C, G, F, D, and A, although the choice of the frequency of the starting point is, of course, somewhat arbitrary. Pythagoras himself decided to stop after he had seven different tones so that his scale consisted of A, B, C, D, E, F, G, not in order of generation but rather arranged in ascending order of frequency.

There is no natural reason to stop here, but the seven-stringed lyre was a well-known instrument in Greece, and evidently this scale had some appeal, being neither too simple nor too complex. But, is there a natural, **mathematical** stopping place in the production of tones by this method?

One such natural stopping place would be if, in continuing to generate new tones at the interval of a fifth above the previous one, one somehow arrived back at one's original tone, thus forming a closed cycle of tones. A look at the mathematics shows us, however, that this can never happen since our multiplier of (3/2) means that any new note will have the frequency of middle C multiplied by $3^n/2^m$, for some integral n and m. The numerator will always be odd and the denominator even. We can never get back to a ratio of 1:1 for any subsequently produced tone of our scale. The Pythagorean method leads to an infinite regress producing new tone after tone, each one at the interval of a fifth from the previous one. We might still ask, however, if we will ever get **near** to the ratio of 1:1, and this time the answer is yes. After producing twelve different tones we arrive at a thirteenth which stands in the ratio of 1.01364:1 to the original middle C. This

95

difference of .01364 is called the "Pythagorean comma."[4] Since, in fact, it is not very large, there seems to be a rationale for stopping after producing twelve different tones. The scale thus produced abounds in intervals of a fifth and is rich enough to produce many other interesting harmonies. Or is it?

Major and Minor Triads

Remember that the underlying theme of the Pythagorean method of generating the musical scale was the production of ratios of 3:2 and 4:3 resulting in tones of such pleasant harmony. But what of other simple ratios, 5:4 or 6:5? In fact, most listeners of music consider tones in these frequency ratios to be very pleasant sounding. They are known in modern notation as the *major third* and *minor third*, respectively. The major *triad*, consisting of C-E-G (in the key of C) is one of the most familiar harmonies in our music. It is composed of two intervals of a third, from C to E (major) and from E to G (minor). However, these thirds do not correspond mathematically to intervals of a third in the Pythagorean scale. For example, the third note above C in the Pythagorean scheme is E which stands in the ratio of 1.2656 to the frequency of C. (See Table I.) This is close to 5:4 (=1.25), but the difference is great enough to be perceptible to the well-trained musical ear. In order to avoid confusion, the interval represented by the ratio 5:4 is called a *just* third as distinct from the Pythagorean third.[5]

---

[4] John Backus, *The Acoustical Foundations of Music*, 2nd ed. (New York, W. W. Norton & Co., 1977), p. 139.

[5] A digression at this point on the relation of harmonics and harmony is perhaps useful. Remember that harmonics are those tones and overtones which occur naturally when one strikes a string which is tuned to a specific frequency. They are all integral multiples of the first frequency, so that the second harmonic has twice the frequency of the first, and is, of course, an octave above it. The third harmonic, three times the frequency of the first, stands in the ratio of 3:2 to the second, and therefore is at an interval of a fifth from this second harmonic. To this point the harmonics are also harmonious. The fourth harmonic, 4 times the frequency of the first, is two octaves above it. By the way, this puts it at an interval of a fourth from the third harmonic. The fifth overtone, in the ratio of 5:4 with the fourth, is therefore a just third above this fourth. Had we begun at C, we would now have C, $_2$C, $_2$G, $_3$C, $_3$E as the first five harmonics, with the $_2$C representing one octave up from the middle C, and $_3$C representing two octaves up, etc. The notes C-E-G, when sounded together, form what is called a major triad, a chord which is one of the fundamental building blocks of traditional western music. Is it a coincidence that these harmonics, naturally occurring as a result of the physical properties of vibrating media, also form the traditionally most pleasing harmonies? In fact, this question is not easy to answer, since not all the harmonics as we continue upward produce such pleasing intervals with each other. After the $_3$E, the next harmonic turns out to be $_3$G, which is, of course, in the ratio of 6:5 to the fifth, and therefore gives the *just minor third*. This too suggests that harmonics sound well together. After that, however, such easy coincidences are not forthcoming. Intervals of 7:6 and 8:7 have much less obvious harmonic appeal. Remember that 9:8 appears as the interval from C to D in the Pythagorean scale, but that was by accident and **not** as part of a construction so that these two notes could be sounded together. The higher harmonics get closer and closer together so that eventually one can find almost any interval represented by harmonics.

It is true, however, that the higher harmonics are generally not as audible as lower ones so that it might be suggested that the first few harmonics play a greater role than others in our perception of harmony. Whether or not this role is as simple as it might appear to be is open to question. It was the theory of Helmholtz over 100 years ago that consonance did not result from the coincidence of harmonics of a single tone. Rather, he said one had

One could, of course, simply adjust the Pythagorean third down a notch. For example, the E could be moved from the ratio of 1.2656 to the ratio of 1.125 with middle C, but this would destroy the perfection of the interval of the fifth between E and A below it, which was Pythagoras's first organizing principle. Apparently, one cannot have everything.

The interval of the minor third above C, not present at all in the original Pythagorean seven tone scale, does exist **approximately** in the twelve tone scale generated by the Pythagorean method as the note D sharp. An approximate minor third which **is** in the Pythagorean scale, the interval from D to F, is, however, in the ratio[6] of 32:27 (=1.185) instead of the ideal 6:5 (=1.20). Again, we are close, but not on the money. We cannot satisfy the musician who wants both perfect fifths and just thirds without some kind of compromise. A scale with one will not have the other.

Equal Temperament

Let us now consider again the twelve tone scale generated by the Pythagorean method. Admittedly this scale is somewhat make-shift. The interval from one of the twelve tones to the next in the modern scale is called a *semitone*. There are, for example, two semitones between C and D, namely, C to C sharp and C sharp to D. C and D have frequencies in the ratio of 9:8. Each of these semitones should be produced by multiplying the frequency of C by the square root of 9/8 (=1.1606) so that to get from C to D one would multiply the frequency of C by $\sqrt{(9/8)}^2$. This is indeed the case when we check the frequencies in Table I. Likewise we get from D to D sharp by multiplying by 1.606, and similarly to E. From E to F is, however, only one semitone in the modern scale and one hemitone in the Pythagorean scale. What is a *hemitone*? The F was determined in advance to be in the ratio of 4:3 with the C. When we compute the interval from E to F in the Pythagorean scale, we find that we multiply the frequency of the E by 1.0535 (that is, 1.3333/1.2656) instead of by 1.606 to get the F. Thus, the hemitone is smaller than the semitone. This too is a problem, especially when one wants to play major triads in key signatures other than C. Obviously the minor third (= three semitones) from D to F cannot be the same as the minor third between F and A flat. A key which depends on the F minor triad will sound differently from one which makes frequent use of the D minor triad. One does not want to be constrained to work in one key only, but one also does not want each key to have its own peculiar sound. In fact, in the pre-Baroque era, keyboard instruments were tuned in such a way that music in different keys did have very different harmonic characteristics.

The solution to this dilemma which is used on most modern keyboard instruments was invented by J. S. Bach and is called the scale of equal temperament. Bach, who himself played instruments of the keyboard type, decided that music in all key signatures should sound equally well even if this meant a sacrifice

---

to consider, when striking G and C together (or C, G, and E together), how the harmonics of each separate tone interacted with the harmonics of the others. This theory is based on the idea that it is necessary to minimize the beats which are produced when sounds of similar but not identical frequencies are juxtaposed. Suffice it to say there may be some connection between harmonics and harmony, but exactly what connection is not beyond dispute.

[6]Backus, p. 141.

of some of the perfect Pythagorean ratios or of just thirds. What he did was to make all semitones equal. To divide the octave into twelve equal parts, one begins at C, multiplies its frequency by the twelfth root of two to get C sharp, multiplies again by the twelfth root of two to get D, and so on, so that by twelve such multiplications one arrives at the octave and begins the process again. The frequency ratios which result from this procedure are presented in Table II where they may be compared to those of the Pythagorean scale.

Table II

Generation of tones by the method of equal temperament

Name for tone	Ratio of frequency to middle C	Pythagorean method
C	1	1
D	$(\sqrt[12]{2})^2 = (1.05946)^2$	
	$= 1.122$	1.125
E	$(1.05946)^4 = 1.260$	1.2656
F	$(1.05946)^5 = 1.335$	1.333
G	$(1.05946)^7 = 1.498$	1.500
A	$(1.05946)^9 = 1.682$	1.6875
B	$(1.05946)^{11} = 1.888$	1.8984
C	$(1.05946)^{12} = 2.000$	2.000

The comparison of the two scales on paper makes them seem reasonably close, but one must **listen** to the difference to decide the matter aesthetically.

Tuning to equal temperament is of primary interest to keyboard instruments or those instruments which play with them. A solo violinist is free to play a Pythagorean scale, a just scale, or whatever combination s/he chooses. Likewise a string quartet or a singing group can adapt the pitch of notes as it goes along, if all members agree in advance on how each interval is to be played or sung. What scales are in fact chosen by such musicians who have a choice is a matter of considerable interest and discussion,[7] and would make an interesting footnote to this article.

We should note, of course, that the entire subject of harmony is under constant revision by modern trends in music. Intervals which were once considered distasteful are now routine, while those once considered exciting have become commonplace. The ratios discovered by Pythagoras are undoubtedly still of great importance, but we have seen that, whereas certain mathematical ratios do reveal pleasant harmonies, there is no one mathematical system for generating a scale which satisfies all criteria.

---

[7]Backus, p. 155 ff.

Play a Cool Pool

Mathematics Used: *Elementary Algebra,*
Similar Triangles

Consider a pool table 42″ × 84″. Suppose one wishes to make the shot illustrated in Figure 1. Let us assume that the angle of "incidence" is equal to the angle of "reflection" and let x denote the distance, in inches, between point P and the point of contact in the lower side as in Figure 2. Clearly $0 < x < 30$.

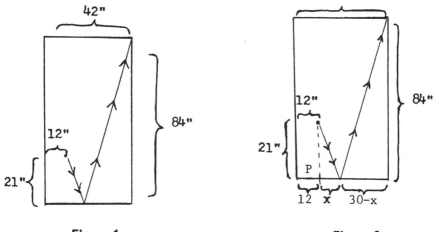

Figure 1   **Figure 2**

From similar triangles, we have the following.

$$\frac{21}{x} = \frac{84}{30-x}$$
$$\frac{1}{x} = \frac{4}{30-x}$$
$$30 - x = 4x$$
$$30 = 5x$$
$$x = 6.$$

Let us consider the same problem from a slightly different point of view. We want to make the same shot as before. However, let us now assume that the ball must move in such a way that the total distance travelled is a minimum (i.e., the shortest distance). To explain further, consider the paths illustrated in Figure 3.

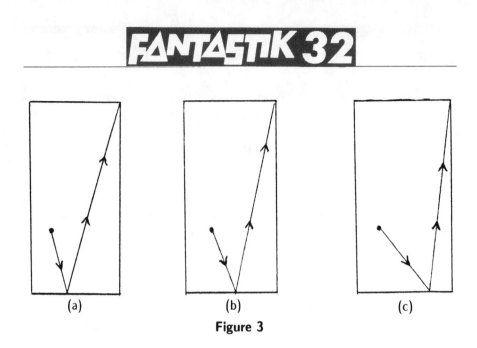

(a)   (b)   (c)

Figure 3

Each of these paths will yield a different "total distance travelled". Our problem is to find which path (not necessarily one of those shown) will yield the shortest route.

Consider a route and part of its mirror image as illustrated in Figure 4.

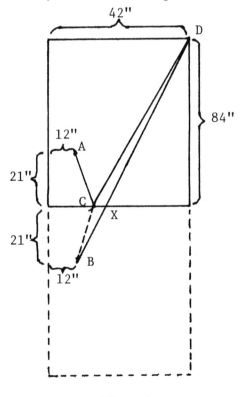

Figure 4

Observe that $\overline{AC} + \overline{CD}$ is the total distance travelled which is equal to the distance $\overline{BC} + \overline{CD}$. Since the sum of two sides of a triangle exceeds the length of the other side, we see that

$$\overline{BC} + \overline{CD} > \overline{BD}.$$

If we could locate the point X as illustrated, then the path A-X-D would yield a distance equal to $\overline{BD}$. Thus, our "best bet" is to locate the point X and take the path A-X-D.

Now, to locate the point X, we consider Figure 5.

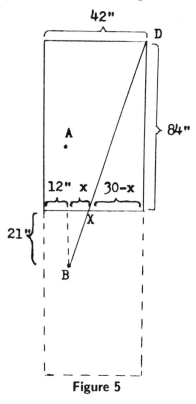

Figure 5

Again, from similar triangles, we have

$$\frac{21}{x} = \frac{84}{30 - x},$$

which is precisely the same equation we solved previously yielding

$$x = 6.$$

From these observations, we can state a useful and remarkable property: IF THE BALL TRAVELS IN SUCH A WAY THAT THE ANGLE OF INCIDENCE IS EQUAL TO THE ANGLE OF REFLECTION, THEN THE BALL WILL TRAVEL IN SUCH A WAY THAT THE TOTAL DISTANCE TRAVELLED IS A MINIMUM, AND *VICE VERSA*.

Practice Problem: Suppose, under either assumption discussed in this example, that one wishes to make the following shot:

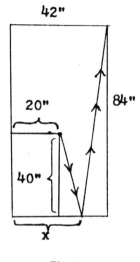

Figure

Determine an appropriate x.

Relatively Relativity

Mathematics Used: *Pythagorean Theorem*

Consider two people, A and B, in two different rooms, denoted RA and RB, respectively. The rooms have glass walls so that A and B can see each other:

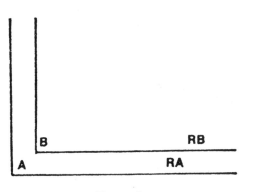

Figure 1

Each person has an expensive, finely tuned watch which can, if necessary, precisely measure fractions of a second. For the sake of demonstration, we are assuming that all measurements can be read accurately to at least 10^{-7}.

In room RB, there is an ingenious device for tracking pulses of light. This device not only emits a light pulse, but also indicates when a pulse is received. A mirror stands 20 m in front of this device.

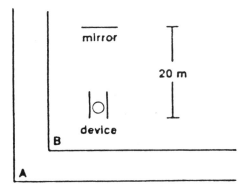

Figure 2

Time, as measured in rooms RA and RB, will be denoted by t_A and t_B, respectively.

Now, suppose light travels at 10 m per second. A pulse emitted by our device is reflected by the mirror, returned, and received by the device as shown here.

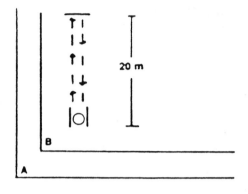

Figure 3

The time elapsed, as measured in either room RA or RB, is

$$t_A = t_B = \frac{40 \text{ m}}{10 \text{ m/sec}} = 4 \text{ sec.}$$

(For this calculation we assume, of course, that rooms RA and RB are stationary relative to each other.)

Suppose now that room RB is moving at the rate of 3 m per second relative to RA.

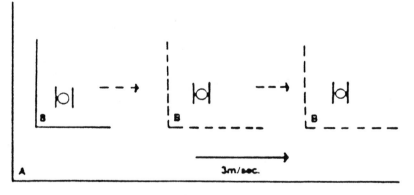

Figure 4

We repeat the previous experiment. The device emits a pulse which is reflected by the mirror and received by the device as indicated. Person B observes

exactly what was observed in the prior experiment, so the time elapsed is still

$$t_B = \frac{40 \text{ m}}{10 \text{ m/sec}} = 4 \text{ sec.}$$

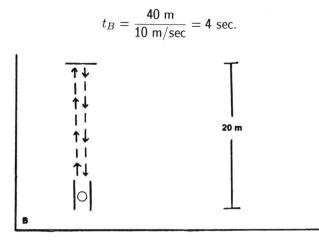

Figure 5

However, person A observes the following:

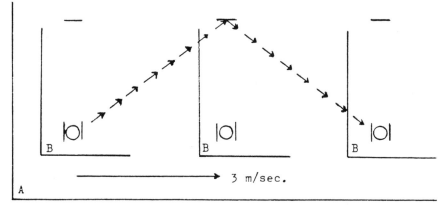

Figure 6

Consider the triangle

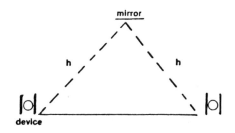

Figure 7

from Figure 6. If t_A is the time elapsed as measured by person A, then from the formula $d = vt$, we see that $d = 3t_A$ so that $3t_A$ is the base of the triangle in Figure 7. In addition, using the assumed speed of light of 10 m/sec and the same formula, $d = vt$, we discover that $2h = 10t_A$ or $h = 5t_A$ where $2h =$ actual distance traveled by the light.

We now have the triangle

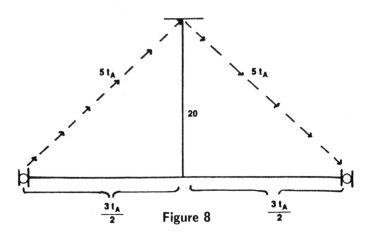

Figure 8

From the Pythagorean theorem,

$$(5t_A)^2 = (3t_A/2)^2 + 400$$
$$(91/4)t_A{}^2 = 400$$
$$t_A{}^2 = 1600/91$$
$$t_A = 4.1931393 \text{ sec.}$$

Therefore, while the time elapsed for person B is **4** sec, the time elapsed for person A is **4.1931393** sec. How can this happen? Einstein's theory accounts for this result by claiming that B's watch "runs more slowly" than A's watch. Transforming the above numbers to a base of one second, we can make the following comparisons:

Table 1

Time, person A	Time, person B
1 sec	0.9539392 sec
1.0482848 sec	1 sec

Suppose now that the speed of light is 80 m/sec. A pulse is emitted by our device, reflected by the mirror, and returned to the device. The time elapsed as measured in either RA or RB, assuming they are stationary relative to each other, is (see Figure 3)

$$t_A = t_B = \frac{40 \text{ m}}{80 \text{ m/sec}} = .5 \text{ sec.}$$

Now suppose that room RB is again moving at the rate of 3 m/sec relative to RA (as in Figure 4). We repeat the previous experiment. The device again emits a pulse which is reflected and returned to the device. Person B observes exactly what was observed in the stationary experiment so the time elapsed is still

$$t_B = \frac{40 \text{ m}}{80 \text{ m/sec}} = .5 \text{ sec.}$$

However, person A again observes the same result shown in Figure 6, yielding the triangle of Figure 7. The base of the triangle remains $3t_A$; but because of the speed of light is now 80 m/sec, we discover

$$2h = 80t_A$$
$$h = 40t_A$$

and the triangle

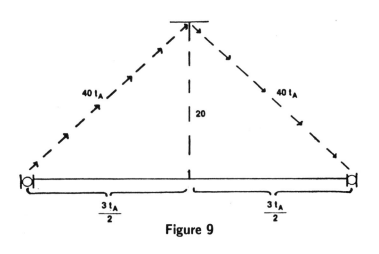

Figure 9

From the Pythagorean theorem

$$(40t_A)^2 = (3t_A/2)^2 + 400$$
$$(6391/4)t_A^2 = 400$$
$$t_A^2 = 1600/6391$$
$$t_A = .5003519 \text{ sec.}$$

Doubling the above figures provides a time comparison chart:

Table 2

Time, person A	Time, person B
1 sec	0.9992967 sec
1.0007038 sec	1 sec

Again, we see that B's watch is "running more slowly" than A's watch. Note, however, that the time differential between A and B is less in Table 2 than that which appears in Table 1. This leads us to hypothesize that if the speed of light were an even larger number, the time differential between A and B would be even smaller (assuming that RB continues to move at the same rate of 3 m/sec relative to A).

Let us again take the speed of light to be 80 m/sec, but suppose that room RB is moving at a speed of 60 m/sec as observed by A. After the experiment is performed, the time elapsed measured by B remains

$$t_B = \frac{40 \text{ m}}{80 \text{ m/sec}} = .5 \text{ sec.}$$

For person A, the triangle corresponding to one in Figure 9 would be

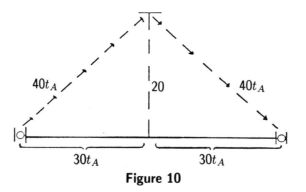

Figure 10

From the Pythagorean theorem

$$(40t_A)^2 = (30t_A)^2 + 400$$
$$700t_A{}^2 = 400$$
$$t_A{}^2 = .5714286$$
$$t_A = .7559289 \text{ sec.}$$

A time comparison chart for this experiment is given in Table 3.

Table 3

Time, person A	Time, person B
1.5118578 sec	1 sec
1 sec	0.6614422 sec

Note that the time differential between A and B has increased from that which appears in Table 1. Indeed, if we expand Table 3 to units of hours and days of a year, we see

Table 4

Time, person A	Time, person B
1 h 31 min	1 h
1 h	40 min
551.83 days	365 days
365 days	241.43 days

Practice Problem 1: Construct additional problems, with appropriate tables, using different speeds for the velocity of light and for the velocity of room RB (relative to RA). What happens if room RB travels, relative to RA, at a speed greater than the assumed speed of light?

Let us generalize the experiment by taking c meters per sec as the speed of light, v meters per sec as the speed of room RB, and L as the distance between the device and the mirror. For observer B, we see the following:

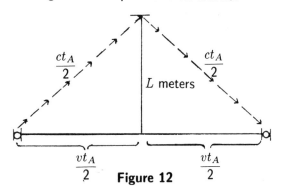

Figure 11

Using $d = vt$,

$$2L = ct_B$$
$$L = \frac{ct_B}{2}. \tag{1}$$

The familiar triangle from A's point of view follows:

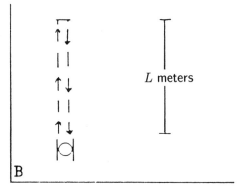

Figure 12

109

Using $L = ct_B/2$ from (1) produces the following:

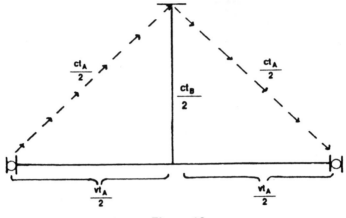

Figure 13

From the Pythagorean theorem

$$(ct_A/2)^2 = (vt_A/2)^2 + (ct_B/2)^2$$
$$[(c/2)^2 - (v/2)^2]t_A{}^2 = (ct_B/2)^2$$
$$t_A = \frac{(c/2)t_B}{\sqrt{(c/2)^2 - (v/2)^2}}.$$

Dividing numerator and denominator by $c/2$ yields

$$t_A = \frac{t_B}{\sqrt{1 - (v^2/c^2)}}. \tag{2}$$

The results show us a relationship between t_A and t_B. Once the latter formula (2) is known, an analysis of the time differential is readily attainable.

Practice Problem 2: In (2), what happens to the time differential between t_A and t_B if **(i)** "v" is close to "c", $v < c$? **(ii)** "v" is very small relative to "c"?

Of Epidemic Proportions

Mathematics Used: *Algebra*

For our initial study of an epidemic (spread of a disease), we consider three pools S, H, and I of people. The S pool is the group of people who are **susceptible** to the disease. The H pool is the group of people who **have** the disease and can communicate it to others. The I pool is the group who are **immune** to the disease (or recovered from the disease).

To be more precise, we let

S_k = number of "susceptibles" k days after initial observation of epidemic.
H_k = number of "haves" k days after initial observation of epidemic.
I_k = number of "immunes" k days after initial observation of epidemic.

Students were asked the following: If you knew the numbers S_k, H_k, and I_k, how would you compute S_{k+1}, H_{k+1}, and I_{k+1}? The following were suggested:

$$S_{k+1} = S_k - .01S_k \tag{1}$$
$$H_{k+1} = H_k - .2H_k + .01S_k \tag{2}$$
$$I_{k+1} = I_k + .2H_k. \tag{3}$$

Reasoning went as follows:

Equation (1): Between day k and day $k+1$, a certain percentage, say 1%, of the "susceptibles" are going to contract the disease and leave this group.

Equation (2): On day $k+1$, we will have the "haves" of the day before (day k) minus those who recover. Assuming a 5 day disease, one could anticipate, in one day, about 1/5 (or .2) of the "haves" recovering and moving to the immune group. In addition, however, we pick up those "susceptibles" who contracted the disease during this day, which we have taken as $.01S_k$.

Equation (3): On day $k+1$, we will have the "immunes" of the day before (day k) plus those who recover from the disease during the day and move into this group (which we have taken as $.2H_k$).

We can rewrite (1), (2), and (3) as

$$\left. \begin{aligned} S_{k+1} &= .99S_k \\ H_{k+1} &= .8H_k + .01S_k \\ I_{k+1} &= I_k + .2H_k. \end{aligned} \right\} \tag{4}$$

Now, suppose our initial data is given by

$$\left. \begin{aligned} S_0 &= 90{,}000 \\ H_0 &= 9{,}000 \\ I_0 &= 1{,}000. \end{aligned} \right\}$$

111

Substituting these numbers into the right hand side of equation (4) we obtain

$$\text{Day 1} \begin{cases} S_1 = 89{,}100 \\ H_1 = 8{,}100 \\ I_1 = 2{,}800 \end{cases}$$

At this point, we went to an electronic calculator and obtained the following:

Day	S_k	H_k	I_k
0	90,000	9,000	1,000
1	89,100	8,100	2,800
2	88,209	7,371	4,420
3	87,327	6,779	5,894
4	86,454	6,296	7,250
5	85,589	5,901	8,509
6	84,733	5,577	9,689
7	83,886	5,309	10,804
8	83,047	5,086	11,866
9	82,217	4,899	12,883
10	81,394	4,733	13,863
11	80,580	4,600	14,810
12	79,775	4,478	15,730
13	78,977	4,372	16,626
14	78,187	4,287	17,500

Table 1

After examining the results, students were asked to criticize the mathematical model we had built. The objections included:

i) We had not taken into account any changes in the total population due to birth or natural death.

ii) We had not considered the possibility that some of the "haves" might die from the disease.

iii) One could not expect that the number of "susceptibles" who contracted the disease during one day would be a **constant** percentage of the number of "susceptibles". Somehow, the number contracting the disease should be related to the number of "susceptibles" **and** the number of "haves" who can communicate the disease.

iv) The model does not consider a possible inoculation program.

The following model for the spread of a disease appeared in *Calculus with Probability* by W. E. Baxter and C. W. Sloyer (Addison-Wesley, 1973).

Consider what happens in the spread of a disease such as measles. In this case, once an individual is affected with the disease, s/he is initially able to infect others. During this initial period s/he is called an *infective*. The individual then moves to a stage where s/he is still affected but no longer an infective, now

112

called an *affective*, and finally to a stage where s/he is immune to the disease. We make the following assumptions:

1. Initially a portion of the population is immune and among new births, a fixed portion δ is immune.
2. There is a daily rate of increase γ in the population through birth.
3. There is a normal daily rate β of attrition through death, equally distributed through the population.
4. The disease causes a daily death rate α among those having the disease.
5. There is a daily rate ω of individuals moving from the infective state to the affective state.
6. There is a daily rate σ of individuals moving from the affective state to the immune state.

Let us assume that there is an **initial** total population of N_0. Let a_0, n_0, and i_0 denote the **initial** number of infectives, affectives, and immunes, respectively. Let N_k denote the total population on the kth day and let a_k, n_k, and i_k denote the number of infectives, affectives, and immunes on the kth day. Making use of the above assumptions, we write

$$N_{k+1} = N_k + (\gamma - \beta)N_k - \alpha(a_k + n_k), \tag{5}$$

$$n_{k+1} = n_k - \beta n_k - \alpha n_k - \sigma n_k + \omega a_k, \tag{6}$$

$$i_{k+1} = i_k - \beta i_k + \sigma n_k + \delta\gamma N_k. \tag{7}$$

Now, a critical step is to assess a_{k+1}, the population of the infectives on the $(k+1)$th day. We have

$$a_{k+1} = a_k - \beta a_k - \alpha a_k - \omega a_k + I \tag{8}$$

where I = the increase in the infective population resulting from infection of individuals in the available population. Previous data suggests I in (8) is proportional to the product of the number of availables and the number of infectives. Let τ denote the proportionality constant. Then

$$a_{k+1} = a_k - \beta a_k - \alpha a_k - \omega a_k + \tau a_k[N_k - (a_k + n_k + i_k)].$$

We have investigated, using an electronic computer, what occurs if we choose

$$N_0 = 1,000,000$$
$$a_0 = 1,000$$
$$n_0 = 1,000$$
$$i_0 = 10,000$$
$$\alpha = 0.001$$
$$\beta = 0.002$$
$$\gamma = 0.004$$
$$\sigma = 0.1$$
$$\omega = 0.1$$
$$\tau = 0.000001.$$

The results are given in Table 2.

Table 2

$$\alpha = 0.00100000 \quad \omega = 0.10000000$$
$$\beta = 0.00200000 \quad \sigma = 0.10000000$$
$$\gamma = 0.00400000 \quad \tau = 0.00000100$$
$$\delta = 0.10000000$$

DAY	TOTAL	AFF & INF	NON-INF	IMMUNE	AVAIL
0	1000000	1000	1000	10000	988000
14	1024773	692140	249011	69295	14327
28	1043379	175566	305956	546976	14881
42	1068342	55068	136162	841764	35348
56	1096792	24990	55729	952099	63974
70	1127046	17665	27133	986882	95366
84	1158445	19136	19530	997004	122775
98	1190703	28097	22257	1003695	136654
112	1223605	43385	32828	1019445	127947
126	1257067	53369	45366	1052000	106332
140	1291291	50721	49807	1096714	94049
154	1326555	44436	46487	1140808	94824
168	1362952	41276	42194	1177982	101500
182	1400443	42184	40710	1210014	107535
196	1438964	45735	42388	1241450	109391
210	1478486	49510	45708	1275845	107423
224	1519040	51683	48642	1314054	104661
238	1560690	52374	50204	1354621	103491
252	1603502	52838	50889	1395781	103994
266	1647507	53931	51645	1436945	104986
280	1692725	55681	52952	1478598	105494
294	1739178	57641	54691	1521476	105370
308	1786895	59448	56503	1565919	105025
322	1835918	61068	58179	1611823	104848
336	1886284	62672	59754	1658972	104886
350	1938033	64391	61364	1707287	104991
364	1991202	66238	63079	1756843	105042
378	2045827	68152	64889	1807764	105022
392	2101949	70091	66749	1860121	104988
406	2159607	72060	68639	1913934	104974
420	2218848	74081	70567	1969218	104982
434	2279711	76167	72550	2026005	104989
448	2342243	78316	74594	2084343	104990

In order to consider a type of analysis which can be made with such models, we return to a somewhat simpler model. We suppose that the population size is a constant N. Again let

S_k = number of "susceptibles" k days after initial observation.
H_k = number of "haves" k days after initial observation.
I_k = number of "immunes" k days after initial observation.

We write

$$S_k = N - (H_k + I_k) \tag{9}$$

$$H_{k+1} = H_k + \beta S_k H_k - \alpha H_k \tag{10}$$

$$I_{k+1} = I_k + \alpha H_k. \tag{11}$$

The student should be able to give the reasoning for each equation.

Note that the computation technique for this model is somewhat different than that of the first model considered. Given N, S_0, H_0, and I_0, we can compute H_1 and I_1. With H_1 and I_1, we compute S_1. Then we compute H_2 and I_2. This enables us to compute S_2, etc.

Let ΔH_k denote the **change** in the number of "haves" from day k to day $k + 1$.

From (10), we see that

$$\Delta H_k = \beta S_k H_k - \alpha H_k. \tag{12}$$

Considering (12), we have

$$\Delta H_k = (\beta S_k - \alpha) H_k.$$

Hence, we have the following:

i) When $(\beta S_k - \alpha) > 0$, the number of "haves" is increasing.
ii) When $(\beta S_k - \alpha) < 0$, the number of "haves" is decreasing.

If the number of "susceptibles" S_k is "large", we could expect $\beta S_k - \alpha$ to be positive, in which case the number of "haves" is increasing. However, if S_k is "small", we could expect $\beta S_k - \alpha$ to be negative, in which case the number of "haves" is decreasing.

The **critical point** for an epidemic is reached when the number of "susceptibles" S is such that $\beta S - \alpha = 0$, or

$$S = \frac{\alpha}{\beta}.$$

We could probably expect the following to happen: We start with a "large" number of "susceptibles", i.e., S_0 is "large". The number of "susceptibles" decreases to S. During this period, the number of "haves" is increasing. The number of "susceptibles" continues to decrease below S, but the number of "haves" is now decreasing.

$$S_0 \; \begin{array}{c} \downarrow \end{array} \left. \begin{array}{c} \\ \\ \end{array} \right\} \begin{array}{l} \text{number of "haves"} \\ \text{is increasing} \end{array}$$

$$S \; \begin{array}{c} \downarrow \end{array} \left. \begin{array}{c} \\ \\ \end{array} \right\} \begin{array}{l} \text{number of "haves"} \\ \text{is decreasing} \end{array}$$

115

We might consider the critical point S to be that point at which the spread of the disease changes from "epidemic proportions" to "nonepidemic proportions". One would like to increase S so that the spread of the disease changes its nature (in terms of the number of "haves") as quickly as possible. Observe that S will increase if we increase α or decrease β. If an inoculation were discovered which was at least partially effective, then we could expect a decrease in β. (Why?) If a medicine were discovered which reduced the time that one "had" the disease (and hence, could communicate the disease), then we could expect an increase in α. (Why?)

Practice Problem 1: For the last model discussed in this section, let ΔS_k denote the **change** in the number of "susceptibles" and ΔI_k the **change** in the number of "immunes" from day k to day $k+1$. Show that ΔS_k is never positive and ΔI_k is never negative.

Practice Problem 2: One group of doctors asks for a one-million-dollar research grant and expects within one year to have an inoculation that will decrease β (in the last model of this section) by 25%. Another group of doctors asks for a one-million-dollar research grant and expects within one year to have a medicine that will increase α by 30%. If only **one** million dollars is available for such research, which group should get the money?

Waste Makes Haste

Mathematics Used: *Calculus, Differential Equations*

Let $W(t)$ denote the amount of waste in a certain lake at time t. It is known that the waste decomposes at a rate proportional to the amount of waste present so that

$$\frac{dW}{dt} = -KW \qquad (1)$$

where $K > 0$ is a constant which depends on the temperature of the water, the chemical composition of the waste, etc. From (1), we have

$$\frac{dW}{W} = -K\,dt$$
$$\int \frac{dW}{W} = \int -K\,dt$$
$$\ln W = -Kt + a$$

where a is a constant, Hence,

$$W = be^{-Kt} \qquad (b = e^a). \qquad (2)$$

Suppose at time $t = 0$ when we begin observing this process, that there are W_0 units of waste in the water. Substituting $t = 0$ in (2), we have

$$W_0 = b$$

and, hence, one has

$$W(t) = W_0 e^{-Kt}.$$

Numerical Calculations: Suppose $K = .35$. If there are originally 1500 pounds of waste in the lake and t is measured in days, then 2 days later, the amount of waste will be

$$W(2) = 1500e^{-.35(2)} = 744.878 \text{ lbs.} \qquad (3)$$

Suppose $K = .35$ and there are originally 1500 pounds of waste in the lake. How many days are required for the amount of waste to be reduced to 400 pounds? Here we must solve

$$400 = 1500e^{-.35t} \qquad (4)$$

for t. From (4), we have

$$\ln 400 = \ln 1500 - .35t$$
$$t = \frac{\ln 1500 - \ln 400}{.35}$$
$$t = 3.776 \text{ days.}$$

Let us now consider the situation where waste is constantly being dumped into the lake at the constant rate of L pounds per day. Then

$$\frac{dW}{dt} = L - KW \tag{5}$$

$$\frac{dW}{dt} + KW = L. \tag{6}$$

Using the standard techniques for solving such equations, we consider

i) the associated homogeneous equation

$$\frac{dW}{dt} + KW = 0$$

with characteristic equation

$$r + K = 0$$

and obtain the solution

$$W_h = ae^{-Kt} \qquad (a \text{ a constant}). \tag{7}$$

ii) a particular solution W_p to (6). We assume

$$W_p = b$$

where b is a constant. Then

$$\frac{dW_p}{dt} = 0$$

and substituting into (6), we have

$$0 + Kb = L$$

$$b = \frac{L}{K}.$$

Thus

$$W_p = \frac{L}{K}. \tag{8}$$

Adding the results in (7) and (8), we obtain

$$W(t) = ae^{-Kt} + \frac{L}{K} \tag{9}$$

as the general solution of (6). Again assuming that $W = W_0$ when $t = 0$, one has

$$W_0 = a + \frac{L}{K}$$

$$a = W_0 - \frac{L}{K}.$$

Substituting this result into (9), we have

$$W(t) = \left(W_0 - \frac{L}{K}\right)e^{-Kt} + \frac{L}{K}. \tag{10}$$

Numerical Calculations: Assume t is measured in days,

$$K = .35,$$
$$W_0 = 1500 \text{ (lbs.)},$$
$$L = 180 \text{ (lbs./day)}.$$

After 2 days, the amount of waste is given by

$$W(2) = \left(1500 - \frac{180}{.35}\right)e^{-(.35)2} + \frac{180}{.35} = 1003.777 \text{ (lbs.)}.$$

We can analyze (10) in the following way: As t gets large, $W(t)$ gets close to L/K. If $(W_0 - L/K) > 0$ or $W_0 > L/K$, then the amount of waste is always larger than L/K, while if $(W_0 - L/K) < L/K$, then the amount of waste is always smaller than L/K.

Numerical Calculations: Assume t is measured in days,

$$K = .35,$$
$$W_0 = 1500 \text{ (lbs.)},$$
$$L = 180 \text{ (lbs./day)}.$$

Then

$$\frac{L}{K} = \frac{180}{.35} = 514.286.$$

Thus, as t gets large, the amount of waste in the lake will get close to 514.286 pounds. Moreover, $W_0 - L/K = 1500 - 514.286 = 985.714 > 0$ and, hence, the amount of waste will always be greater than 514.286 pounds. Suppose we wish to know how many days will be required for the amount of waste to be reduced to 600 pounds. From (10), we must solve the equation

$$600 = \left(1500 - \frac{180}{.35}\right)e^{-(.35)t} + \frac{180}{.35}$$

for t. Hence, we have

$$85.714 = 985.714e^{-.35t}$$
$$\ln 85.714 = \ln 985.714 - .35t$$
$$t = \frac{\ln 985.714 - \ln 85.714}{.35}$$
$$t = 6.978 \text{ days}.$$

Practice Problem 1: Suppose $K = .31$. If there are originally 1800 pounds of waste in the lake and t is measured in days, how much waste will be present 3 days later?

Practice Problem 2: Suppose $K = .31$ and there are initially 1800 pounds of waste in the lake. Suppose 100 pounds of waste are dumped into the lake every day. What happens as t gets large? How many days will be required for the amount of waste to be reduced to 600 pounds?

Bend The Trend or Raise The Razor

Mathematics Used: *Algebra, Elementary Probability, Matrix Algebra*

Suppose that there are two brands of razor blades, say brand A and B, which dominate the market. We assume that an individual purchases a new pack of razor blades about once a month and that the following are in effect:

  i) If a person uses brand A this month, there is a probability of 6/10 that s/he will use brand A next month and a probability of 4/10 that s/he will use brand B next month.

  ii) If a person uses brand B this month, there is a probability of 7/10 that s/he will use brand B next month and a probability of 3/10 that s/he will use brand A next month.

We put this information into matrix form as follows:

$$\text{(This Month)} \begin{array}{cc} & \text{(Next Month)} \\ & \begin{array}{cc} A & B \end{array} \\ \begin{array}{c} A \\ B \end{array} & \begin{pmatrix} .6 & .4 \\ .3 & .7 \end{pmatrix} \end{array}$$

We write

$$P_1 = \begin{pmatrix} .6 & .4 \\ .3 & .7 \end{pmatrix}$$

and call P_1 a "one-stage transition matrix".

Consider the probability that an individual now using brand A will be using brand A two months hence. The following "tree" illustrates the possibilities:

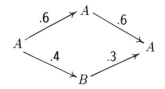

Thus, the desired probability is given by

$$(.6)(.6) + (.4)(.3) = .48.$$

Suppose an individual now uses brand B and we wish to find the probability that s/he will be using brand A two months hence. The following tree illustrates the possibilities:

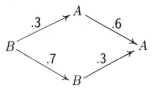

120

Thus, the desired probability is given by

$$(.3)(.6) + (.7)(.3) = .39.$$

Continuing in the above fashion, we form a "two-stage transition matrix".

$$P_2 = \begin{matrix} & A & B \\ (A) & \begin{pmatrix} .48 & .52 \\ (B) & .39 & .61 \end{pmatrix} \end{matrix}$$

A three-stage transition matrix is given by:

$$P_3 = \begin{matrix} & A & B \\ (A) & \begin{pmatrix} .444 & .556 \\ (B) & .417 & .583 \end{pmatrix} \end{matrix}$$

(The student should verify at least one of the entries.)
    An interesting observation is that

$$P_2 = P_1^2 = \begin{pmatrix} .6 & .4 \\ .3 & .7 \end{pmatrix} \begin{pmatrix} .6 & .4 \\ .3 & .7 \end{pmatrix} = \begin{pmatrix} .48 & .52 \\ .39 & .61 \end{pmatrix}$$

and

$$P_3 = P_1^3 = P_1^2 P_1 = \begin{pmatrix} .48 & .52 \\ .39 & .61 \end{pmatrix} \begin{pmatrix} .6 & .4 \\ .3 & .7 \end{pmatrix} = \begin{pmatrix} .444 & .556 \\ .417 & .583 \end{pmatrix}$$

Indeed, in general, for any positive integer k,

$$P_k = P_1^k.$$

Observe, also, that the sum of the entries in any row of a transition matrix must equal one.
    Suppose at the time one begins observing this activity, that brand A sells to 3/4 of the market and brand B sells to only 1/4 of the market. We use vector notation to indicate this information by writing

$$\bar{x}_0 = (3/4, 1/4)$$
$$\bar{x}_0 = (.75, .25).$$

We now wish to write a similar vector $\bar{x}_1 = (\_,\_)$ where the entries indicate the part of the market which each brand controls one month later. Consider the following:

Now	One Month Later	Probability
	A	.45
.75 A		
	B	.30
	A	.075
.25 B		
	B	.175

Thus, we have
$$\bar{x}_1 = (.525, .475).$$

In a similar fashion, (the student should make the calculations) we have

$$\bar{x}_2 = (.4575, .5425).$$

Another interesting observation is that

$$\bar{x}_1 = \bar{x}_0 P_1 = (.75, .25) \begin{pmatrix} .6 & .4 \\ .3 & .7 \end{pmatrix} = (.525, .475)$$

and

$$\bar{x}_2 = \bar{x}_1 P_1 = \bar{x}_0 P_1^2 = (.75, .25) \begin{pmatrix} .444 & .556 \\ .417 & .583 \end{pmatrix} = (.4575, .5425).$$

That is,
$$\bar{x}_1 = \bar{x}_0 P_1,$$
$$\bar{x}_2 = \bar{x}_0 P_1^2.$$

Similarly,
$$\bar{x}_3 = \bar{x}_0 P_1^3$$

and, in general,
$$\bar{x}_k = \bar{x}_0 P_1^k.$$

The vectors
$$\bar{x}_0 = (.75, .25)$$
$$\bar{x}_1 = (.525, .475)$$
$$\bar{x}_2 = (.4575, .5425)$$
$$\vdots$$
etc.

give one some indication of the "trends" which can be expected to take place.

There is an important theorem (known as the **First Ergodic Theorem**) which states that if one begins with a matrix P_1 which has no zero entries, then

   i) there exists a unique vector $\bar{x}$ such that $\bar{x}P_1 = \bar{x}$ ($\bar{x}$ is called a **fixed vector** for P_1).

   ii) as k gets large, P_k ($= P_1^k$) gets close to a matrix P where every row of P is $\bar{x}$.

   iii) for any initial vector $\bar{x}_0$, as k gets large, $\bar{x}_k$ gets close to $\bar{x}$.

Relative to the above theorem, an important calculation is that of the vector $\bar{x}$. Now, for the example under consideration, suppose

$$\bar{x} = (a, b).$$

Then, if $\overline{x}P_1 = \overline{x}$, we must have

$$(a, b) \begin{pmatrix} .6 & .4 \\ .3 & .7 \end{pmatrix} = (a, b)$$

$$(.6a + .3b, .4a + .7b) = (a, b)$$

$$\left. \begin{array}{r} .6a + .3b = a \\ .4a + .7b = b \end{array} \right\}$$

$$\left. \begin{array}{r} -.4a + .3b = 0 \\ .4a + (-.3b) = 0 \end{array} \right\}$$

or simply

$$.4a - .3b = 0.$$

Recalling that $a + b = 1$, we must solve the equations

$$\left. \begin{array}{r} .4a - .3b = 0 \\ a + b = 1 \end{array} \right\}$$

which yield

$$a = 3/7, \qquad b = 4/7.$$

That is,

$$\overline{x} = (3/7, 4/7).$$

The importance of this result is the following: No matter what is the present status of the market, if P_1 continues as the "shift" matrix, then the trend will be toward the situation where brand A controls 3/7 of the market and brand B controls 4/7 of the market.

Activities such as the one discussed in this section are called **Markov Processes** or **Markov Chains**.

Practice Problem: For each of the following one-stage transition matrices, find P_2 and P_3. Given $\overline{x}_0 = (1/3, 2/3)$, find $\overline{x}_1$ and $\overline{x}_2$. Determine the vector $\overline{x}$ and interpret.

i) $\begin{pmatrix} .2 & .8 \\ .7 & .3 \end{pmatrix}$ ii) $\begin{pmatrix} .6 & .4 \\ .1 & .9 \end{pmatrix}$

A Chop Drop

Mathematics Used: *Distance Formula*

When a patient arrives in an intensive care unit or a trauma center, numerous tests and evaluations lead to a large accumulation of data. A problem that physicians have faced is the following: How does one turn this data into information that can be useful in terms of medical action? Teams of physicians and mathematicians have devoted considerable time and effort to distilling, from as many as sixty different physiological and biochemical variables, those that contain the most useful information. For some purposes, the following four variables have survived:

1. **Serum Creatinine (C)**—a measure of the kidney functions. A high level of creatinine indicates a kidney dysfunction. The unit is milligrams percent. The approximate range is 0–25.
2. **Hematocrit (H)**—the percent of the blood that is composed of red blood cells. The red blood cells carry oxygen, and a low reading of hematocrit is usually associated with respiratory or circulatory problems. The unit is percent. The approximate range is 0–65.
3. **Serum Osmolality (O)**—a measure of the number of particles in the blood. High osmolality levels cause body cells to lose fluids and become dehydrated, with bad effects for various bodily organs. The unit is milliosmoles per kilogram of water. The approximate range is 260–400.
4. **Systolic Blood Pressure (P)**—the pressure exerted by the heart and arteries to keep the blood circulating in the blood vessels throughout the body. The unit is milligrams of mercury. The approximate range is 0–280.

From the letter symbol used for each of these variables, one observes a reason for the medical acronym CHOP. This fantastik will demonstrate how the familiar distance formula can be used to give a quantitative meaning to CHOP as a medical index.

Consider just the two variables P (systolic blood pressure) and C (serum creatinine). For healthy adults, the average value of P is 127.0 mm Hg (millimeters of mercury). We write

$$\overline{P} = 127.0$$

The following formula is used to denote the "difference from normal":

$$P_D = |P - \overline{P}| = |P - 127.0|$$

Thus, for example, if the blood pressure of an individual is measured as $P = 135$, then

$$P_D = |135 - 127| = 8.$$

In a similar fashion, for the variable C, $\overline{C} = 1.0$ and

$$C_D = |C - \overline{C}| = |C - 1.0|.$$

Consider a patient with $P = 130$ and $C = 5$. For this patient $P_D = 3$ and $C_D = 4$. If one plots the point $(3, 4)$ in a $P_D C_D$-plane, one easily sees that the distance between this point and the origin is $5 = \sqrt{3^2 + 4^2}$. The origin, of course, corresponds to a perfectly normal (healthy) individual. The greater the distance between the point (P_C, C_D) and the origin, the farther the patient is from normal. One could use the formula

$$N = \sqrt{P_D^2 + C_D^2}$$

as a measure of how close a patient is to normal.

Consider a patient that is being "tracked" for two consecutive days (a physician is looking for the effect of a certain medical treatment).

Example 1:

Day	P	C
1	120	2.0
2	143	3.0

On the first day, $P_D = 7$ and $C_D = 1$, so $N = \sqrt{50} \approx 7.07$, whereas on the second day, $P_D = 16$ and $C_D = 2$, so $N = \sqrt{260} \approx 16.12$. Thus, this patient is farther from normal on the second day than on the first day.

Consider another patient with the following data:

Example 2:

Day	P	C
1	149	3.0
2	138	2.0

On the first day, $N \approx 22.09$ and on the second day, $N \approx 11.05$. Thus, this patient is closer to normal, in the sense under discussion, on the second day compared to the first day.

As the reader may have surmised, using N as a measure of "distance from normal", presents a problem. That is, problems are encountered in using N as a source of medical information. Since the variable P is a "large" number and C is a "small" number, N is more influenced by P. For example, consider the following data:

Example 3:

Day	P	C
1	147	2.0
2	140	4.0

On day one, $N \approx 20.02$, whereas on the second day, $N \approx 13.34$. However, the creatinine level of 4.0 on the second day is a much more serious symptom than the blood pressure level of 147 on the first day. In a medical sense, this patient is farther from normal on the second day compared to the first day.

One can mitigate this problem by "normalizing" the variables. If X is a variable with average $\overline{X}$, then the normalized value of X, denoted X_n, is given by

$$X_n = \frac{X - \overline{X}}{S(X)}$$

where $S(X)$ denotes the standard deviation. For the variables P and C, it has been established that $S(P) = 21$, whereas $S(C) = 0.5$. Hence,

$$P_n = \frac{P - \overline{P}}{S(P)} = \frac{P - 127}{21}$$

$$C_n = \frac{C - \overline{C}}{S(C)} = \frac{C - 1.0}{0.5}.$$

For the patient in Example 3, on day 1, $P_n = 0.95$ and $C_n = 2$. If one plots the point $(0.62, 6)$ in a $P_n C_n$-plane, then the distance N_n between this point and the origin is given by

$$N_n = \sqrt{0.95^2 + 2^2} \approx 2.21.$$

On day 2,

$$N_n = \sqrt{0.62^2 + 6^2} \approx 6.03.$$

Thus, in this new and more meaningful sense, this patient is farther from normal on the second day than the first day.

The CHOP index, now used in medical science, is an extension of N to four-dimensional space, including all the variables $C, H, O,$ and P. That is,

$$\text{CHOP index} = \sqrt{C_n^2 + H_n^2 + O_n^2 + P_n^2}$$

which is the distance between the point (C_n, H_n, O_n, P_n) and the origin in CHOP-space.

Practice Problem: Using $\overline{H} = 37.0$, $S(H) = 6.0$, $\overline{O} = 292.0$, and $S(0) = 15.0$, compute the CHOP Index for the following patients:

Patient	C	H	O	P	CHOP index
1	1.0	37.0	292.0	127.0	_____
2	3.0	25.0	322.0	64.0	_____
3	0.5	43.0	307.0	106.0	_____
4	6.0	13.0	367.0	43.0	_____

Based on the CHOP Index, which patient is in i) the best condition, ii) the worst condition?

A Smug Drug Revisited

Mathematics Used: *Difference Equations (See Appendix II),*
Review Fantastik 16

A doctor determines that there are 10 units of a certain drug in a body. At the end of any hour, assuming no injections of the drug are given, there is one-third of the active concentrate that existed in the body at the beginning of the hour. The rest is removed as waste or made inactive by chemical reactions in the body. Injections of 10 units of the drug are made every 24 hours. Let B_k and A_k denote the amounts of active concentrate in the body before and after, respectively, the kth injection. Then

$$B_{k+1} = (1/3)^{24} A_k \tag{1}$$

and

$$A_k = 10 + B_k. \tag{2}$$

Substituting (2) into (1), we obtain the difference equation

$$B_{k+1} = (1/3)^{24} 10 + (1/3)^{24} B_k. \tag{3}$$

Using the usual technique for solving such equations, we perform the following:

i) Rewrite (3) as

$$B_{k+1} - (1/3)^{24} B_k = (1/3)^{24} 10. \tag{4}$$

ii) Consider the associated homogeneous equation

$$B_{k+1} - (1/3)^{24} B_k = 0$$
$$B_{k+1} = (1/3)^{24} B_k.$$

Assuming $B_0 = a$ (a constant), we have

$$B_1 = (1/3)^{24} a$$
$$B_2 = (1/3)^{2 \cdot 24} a$$
$$B_3 = (1/3)^{3 \cdot 24} a$$

and, in general,

$$B_k = (1/3)^{k \cdot 24} a.$$

iii) Assume that B_k is a constant number L. Then substituting into (4), we have

$$L - (1/3)^{24} L = (1/3)^{24} 10$$
$$L = \frac{(1/3)^{24} 10}{1 - (1/3)^{24}}.$$

(The student should perform the necessary algebra.)
iv) Write the general solution of (3):

$$B_k = (1/3)^{k \cdot 24} a + \frac{(1/3)^{24} 10}{1 - (1/3)^{24}}.$$

Letting $k = 0$, and assuming $B_0 = 0$, we have

$$0 = a + \frac{(1/3)^{24} 10}{1 - (1/3)^{24}}$$

$$a = -\frac{(1/3)^{24} 10}{1 - (1/3)^{24}}.$$

Thus,

$$B_k = (1/3)^{k \cdot 24} \left[-\frac{(1/3)^{24} 10}{1 - (1/3)^{24}} \right] + \frac{(1/3)^{24} 10}{1 - (1/3)^{24}}$$

$$B_k = \frac{(1/3)^{24} 10 [1 - (1/3)^{k \cdot 24}]}{1 - (1/3)^{24}}.$$

(The student should perform the necessary algebra.) Hence,

$$A_k = 10 + B_k$$

$$= 10 + \frac{(1/3)^{24} 10 [1 - (1/3)^{k \cdot 24}]}{1 - (1/3)^{24}}$$

$$= \frac{10 [1 - (1/3)^{(k+1) \cdot 24}]}{1 - (1/3)^{24}}.$$

(The student should perform the necessary algebra.) Thus, we have

$$B_k = \frac{(1/3)^{24} 10 [1 - (1/3)^{k \cdot 24}]}{1 - (1/3)^{24}}$$

and

$$A_k = \frac{10 [1 - (1/3)^{(k+1) \cdot 24}]}{1 - (1/3)^{24}}.$$

Now, as k gets large, $(1/3)^{k \cdot 24}$ and $(1/3)^{(k+1) \cdot 24}$ get close to zero. Hence, B_k and A_k get close, respectively, to B and A given by

$$B = \frac{(1/3)^{24} 10}{1 - (1/3)^{24}}$$

and

$$A = \frac{10}{1 - (1/3)^{24}}.$$

Suppose, at some point in time, that the active concentrate in the body before an injection actually reaches B. Then, immediately after the injection there are

$$10 + B = 10 + \frac{(1/3)^{24} 10}{1 - (1/3)^{24}} = \frac{10}{1 - (1/3)^{24}} = A$$

units of active concentrate in the body. Before the next injection 24 hours later, there are

$$(1/3)^{24} A = \frac{(1/3)^{24} 10}{1 - (1/3)^{24}} = B$$

units of active concentrate in the body.

Practice Problem 1:   Suppose that 6 units of a certain drug are now in a body. At the end of any hour, assuming no injections of the drug are given, there is one-half of the active concentrate that existed in the body at the beginning of the hour. Suppose injections of 6 units are given every 9 hours. Let B_k and A_k denote the amount of active concentrate in the body **before** and **after**, respectively, the kth injection. Use difference equation techniques to find compact expressions for B_k and A_k. Find A and B. Show that if the active concentrate in the body reaches B before an injection, then immediately after the next injection, there are B units of active concentrate in the body.

Practice Problem 2:   Generalize the problem of this section in the following way: Suppose that n units of a certain drug are now in the body. At the end of any hour, assuming no injections of the drug are given, there is a part r, $0 < r < 1$, of the active concentrate that existed in the body at the beginning of the hour. Suppose injections of n units are given every t hours. Let B_k and A_k denote the amount of active concentrate in the body **before** and **after**, respectively, the kth injection. Use difference equation techniques to find compact expressions for B_k and A_k. Find A and B. Show that if the active concentrate in the body reaches B before an injection, then immediately after the injection there are A units in the body. Show that before the next injection there are B units of active concentrate in the body.

A Lank Tank

Mathematics Used: *Difference Equations (See Appendix II)*

An aquarium contains 6 gallons of water. Each week one gallon evaporates and a fresh gallon is poured into the tank. Suppose that one gallon of fresh water contains .01 gallon of salt. (The number .01 is used for pedagogical reasons. The salinity of "fresh" water is actually less.) When water evaporates, the salt remains. Thus, one begins with .06 gallons of salt in the tank. A week later, after a fresh gallon is poured into the tank, the salt content is .07. One week later, after another fresh gallon is poured into the tank the salt content is .08, etc. We thus obtain the following table:

Week	0	1	2	3	4	5	6	$\cdots$
Salt Content	.06	.07	.08	.09	.10	.11	.12	$\cdots$

Hence, the salt content is constantly increasing. Eventually, the salt content will be so great that fish will be unable to live in the tank.

Suppose now that we begin with the same tank having six gallons of fresh water and hence .06 gallons of salt. One week later, there are five gallons of water left. One gallon is removed and two fresh gallons are poured into the tank. Assuming that the salt is uniformly distributed throughout the tank, the salt content is now

$$(.06) - \frac{1}{5}(.06) + .02 = (.8)(.06) + .02 = .068.$$

After another week, we again remove one gallon and pour two fresh gallons into the tank. The salt content is now

$$(.068) - \frac{1}{5}(.068) + .02 = (.8)(.068) + .02 = .0744.$$

Assuming that this process continues, the student should be able to show that after another week the salt content is .07952 gallons. We thus construct the following table:

Week	0	1	2	3	$\cdots$
Salt Content	.06	.068	.0744	.07952	$\cdots$

The student should try to predict what is going to happen to the salt content in the tank.

When asked for predictions, students responded with the following:

  i) The salt content is going to keep increasing but at a slower rate than in the case where we simply pour one fresh gallon into the tank each week. (See Figure 1.)

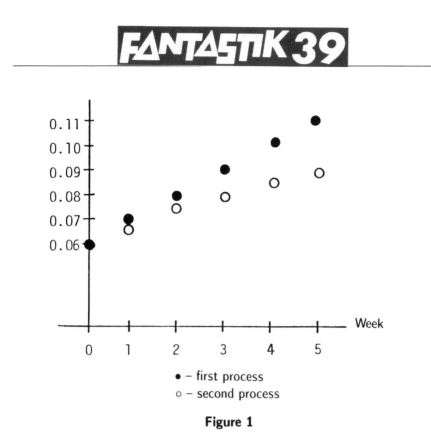

• – first process
o – second process

Figure 1

ii) Eventually, there will be so much salt in the water that at the end of a week, when one removes a gallon of water (in addition to the gallon which has evaporated), the amount of salt removed will be greater than that which is inserted with the two gallons of fresh water. Thus, the salt content will begin to fluctuate as shown in the figure below.

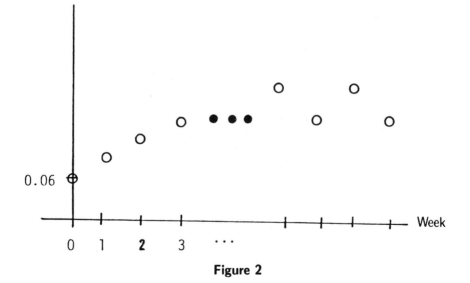

Figure 2

One student observed that if the salt content ever reached .1 gallon, then it "forever" would remain at .1 gallon.

Let us consider this problem from the following point of view: Suppose x_k is the salt content in the tank at the end of week k, after the removing and pouring activities take place. Then

$$x_{k+1} = x_k - \frac{1}{5}x_k + .02$$
$$x_{k+1} = .8x_k + .02$$

where $x_0 = .06$.

We used a microcomputer to partially solve this equation and the following table resulted:

Week	Salt Content
0	.06
1	.068
2	.0744
3	.0795
4	.0836
5	.0869
6	.0895
7	.0916
8	.0933
9	.0946
10	.0957
11	.0966
12	.0973
13	.0978
14	.0982
15	.0986
16	.0989
17	.0991
18	.0993
19	.0994
20	.0995

The pattern was now rather obvious. The salt content does keep increasing but does not get above .1. The salt content continues to get closer and closer to .1.

Now, equation (1) is a **difference equation** and this equation may be solved by the following steps:

i) Rewrite the equation as

$$x_{k+1} - .8x_k = .02. \tag{2}$$

ii) Consider the associated homogeneous equation

$$x_{k+1} - .8x_k = 0$$
$$x_{k+1} = .8x_k.$$

Assuming, for this equation, that $x_0 = A$, then

$$x_1 = .8A$$
$$x_2 = (.8)^2 A$$
$$x_3 = (.8)^3 A$$

and, in general,

$$x_k = (.8)^k A.$$

iii) Assume that x_k is a constant number L. Substituting into (2), we have

$$L - .8L = .02.$$
$$.2L = .02$$
$$L = .1.$$

iv) Using the sum of the results in (ii) and (iii), we write

$$x_k = (.8)^k A + .1.$$

Letting $k = 0$, we have

$$.06 = A + .1$$
$$A = -.04.$$

Thus,

$$x_k = (.8)^k (-.04) + .1.$$

We thus see the following:

a) The salt content is always less than .1 gallon.
b) As k gets larger and larger, $(.8)^k$ gets closer and closer to zero and, hence, the salt content gets closer and closer to .1.

Practice Problem 1: Suppose that at the end of each week, one removes two gallons of water and inserts 3 fresh gallons. Set up and solve a difference equation to find an expression for the salt content at the end of any given week. Analyze your result.

Practice Problem 2: Suppose that fish will die if the salt content exceeds .09 gallons. How much water must be removed at the end of each week, with an appropriate amount inserted, to keep the fish alive?

Practice Problem 3: Suppose that fish will die if the salt content exceeds 2 gallons. How much water must be removed at the end of each week, with an appropriate amount inserted, to keep the fish alive?

Cobwebs

Mathematics Used: *Difference Equations*
(See Appendix II)

In this section, we consider an economic model related to supply and demand.

Each year a certain farmer grows corn as a commercial crop. There is no carry-over of this crop from year to year. A decision on how much corn will be planted is based on the price from the previous year. If the price is high, more is planted, while if it is low, less is planted. During the year, the demand for the crop is dependent upon the price at the time of sale. As the price goes up, the demand goes down.

Therefore, if we let p_n be the price in the nth year, then s_{n+1} (the supply in the $(n+1)$st year) is a function of p_n, while d_{n+1} (the demand in the $(n+1)$st year) is a function of p_{n+1}.

Assuming market price is determined by equilibrium between supply and demand, we seek a price so that $s_{n+1} = d_{n+1}$.

We assume that

$$s_{n+1} = ap_n - b$$
$$d_{n+1} = -cp_{n+1} + d,$$

where $a, b, c,$ and d are positive real numbers.

We wish to discuss the behavior of the prices $p_1, p_2, p_3, \ldots$ in subsequent years, given an initial (present) price p_0.

The equation $s_{n+1} = d_{n+1}$ and the equations above yield

$$ap_n - b = -cp_{n+1} + d$$
$$p_{n+1} = -\frac{a}{c}p_n + \left(\frac{b}{c} + \frac{d}{c}\right)$$

or, with the obvious substitutions,

$$p_{n+1} = -Ap_n + B, \qquad A > 0, \ B > 0,$$
$$p_{n+1} + Ap_n = B. \tag{1}$$

Using the usual techniques, we seek a solution by looking first at the associated homogeneous equation

$$p_{n+1} + Ap_n = 0. \tag{2}$$

A solution of (2) is given by

$$p_n = C(-A)^n,$$

where C is constant.

134

Now, a particular solution to (1) is found by setting $p_n = D$ (a constant) for all n. Then, substituting into (1), we have $D + AD = B$ or $D = B/(1 + A)$. Therefore, the general solution of (1) is given by

$$p_n = C(-A)^n + \frac{B}{1 + A}. \tag{3}$$

Now, we are given p_0. Hence, substituting $n = 0$ into (3), we have

$$p_0 = C(-A)^0 + \frac{B}{1 + A}$$
$$p_0 = C + \frac{B}{1 + A};$$
$$C = p_0 - \left[\frac{B}{1 + A}\right].$$

Substituting in (3), we have

$$p_n = p_0(-A)^n + \left(\frac{B}{1 + A}\right)[1 - (-A)^n]. \tag{4}$$

We consider the result in (4) for three important ranges of values of A.

(1) When $0 < A < 1$, then $(-A)^n$ gets close to zero as n gets large and hence p_n gets close to $B/(1 + A)$. The significance of this result can be seen geometrically if we graph d_n and s_{n+1} as functions of the price p_n (Figure 1).

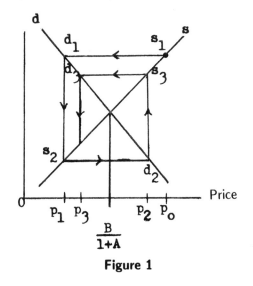

Figure 1

We start with p_0. The point on the s-line above p_0 gives s_1. Moving horizontally, we find d_1 ($d_1 = s_1$). The price "under" d_1 is p_1 which in turn yields s_2. Moving horizontally again, we find d_2 ($s_2 = d_2$) and continue this process. Thus, we see our "cobweb" tighten about the price value $B/(1 + A)$ and the corresponding intersection on the graph.

(2) Suppose $A = 1$. Then equation (4) becomes

$$p_n = (-1)^n p_0 + \frac{B}{2}[1 - (-1)^n].$$ (5)

Hence,

$$p_1 = -p_0 + B.$$

Also,

$$p_3 = -p_0 + B$$

and, indeed,

$$-p_0 + B = p_1 = p_3 = p_5 = \cdots.$$

From (5), we see that

$$p_2 = p_0, \qquad p_4 = p_0,$$

and, indeed,

$$p_0 = p_2 = p_4 = p_6 = \cdots.$$

The situation is illustrated geometrically in Figure 2.

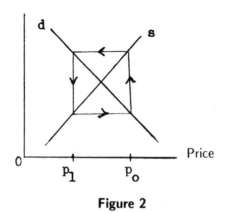

Figure 2

(3) Suppose $A > 1$. From equation (4), we see that p_n will get large in the positive and negative sense as n gets large. We illustrate this process in Figure 3.

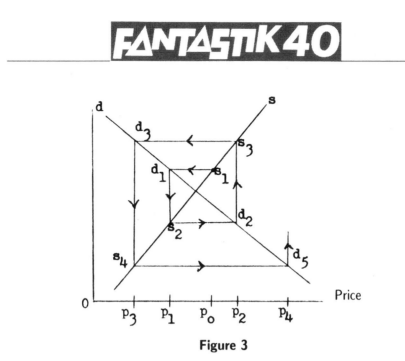

Figure 3

Some final remarks regarding this model. Observe that A is the ratio of the slopes of the straight lines giving supply and demand, respectively. Also we should keep in mind the linear behavior which was assumed in this model. A more sophisticated model, which did not make such assumptions, would be most difficult to handle. Indeed, such problems are frequently incapable of explicit solution.

Practice Problem: Follow the details of this section with

$$s_{n+1} = 3p_n - 2$$

and

$$d_{n+1} = -6p_{n+1} + 8.$$

Appendix I

Arithmetic-Geometric Mean Inequality

We begin with the basic *arithmetic-geometric mean inequality* which states that

$$\frac{a+b}{2} \geq \sqrt{ab} \tag{1}$$

where $a, b \geq 0$. Moreover, equality holds in (1) if and only if $a = b$. This result can easily be proved by starting with the known inequality

$$(a-b)^2 \geq 0. \tag{2}$$

Thus,

$$a^2 - 2ab + b^2 \geq 0$$

or (adding $4ab$ to both sides)

$$a^2 + 2ab + b^2 \geq 4ab$$
$$\frac{(a+b)^2}{4} \geq ab.$$

Assuming $a, b \geq 0$, we arrive (by taking the square root of each side) at

$$\frac{a+b}{2} \geq \sqrt{ab}. \tag{3}$$

Moreover, equality holds in (2) and, hence, in (3) if and only if $a = b$.

One can show that

$$\frac{a_1 + a_2 + a_3}{3} \geq \sqrt[3]{a_1 a_2 a_3} \tag{4}$$

where $a_1, a_2, a_3 \geq 0$. Moreover, equality holds in (4) if and only if $a_1 = a_2 = a_3$.

The above process generalizes to

$$\frac{a_1 + a_2 + \cdots + a_n}{n} \geq \sqrt[n]{a_1 a_2 \cdots a_n} \tag{5}$$

where $a_i \geq 0$ for $i = 1, 2, \ldots, n$. Moreover, equality holds in (5) if and only if $a_i = a_j$ for each i and j. A proof of (5) can be found in *Algebra and Its Applications* by C. W. Sloyer, published by Addison-Wesley, 1970.

Appendix II

Linear Difference Equations

We consider equations of the form

$$x_{n+1} = ax_n + b \tag{1}$$

where a and b are constants. Such an equation is called a *difference equation*. A solution is simply a sequence $x_0, x_1, x_2, \ldots, x_n, \ldots$ where $x_{k+1} = ax_k + b$ for $k = 0, 1, 2, 3, \ldots$. We shall illustrate, via an example, a method for finding a solution to such equations. We shall then show that this method works for any such equation.

Example: Consider the equation

$$x_{n+1} = 2x_n + 3 \tag{2}$$

with $x_0 = 4$. The equation

$$x_{n+1} - 2x_n = 0$$
$$x_{n+1} = 2x_n \tag{3}$$

is called the *associated homogeneous equation*. Suppose $x_0 = A$ (a constant) in (3). Then

$$x_1 = 2A$$
$$x_2 = 2^2 A$$
$$x_3 = 2^3 A$$

and, in general,

$$x_n = 2^n A. \tag{4}$$

We now suppose that in (2), $x_n = B$ for every n. Substituting into (2), we have

$$B = 2B + 3$$
$$B = -3.$$

That is,

$$x_n = -3 \tag{5}$$

is a *particular solution* of (2). A *general solution* of the difference equation in (2) is now given by adding the results in (4) and (5), or

$$x_n = 2^n A - 3. \tag{6}$$

In our example, we are given that $x_0 = 4$. Substituting $n = 0$ into (6), we have

$$4 = A - 3$$
$$A = 7.$$

Thus, a solution to (2) with $x_0 = 4$ is given by

$$x_n = (2^n)7 - 3. \tag{7}$$

More informally, a solution is given by the sequence

$$4, 11, 25, 53, 109, \ldots.$$

We shall now show that the method described in the above example will work in general. Consider the difference equation

$$x_{n+1} = ax_n + b \tag{8}$$

with x_0 given. Note that if $a = 1$, then a solution sequence is provided by the simple arithmetic sequence

$$x_0, x_0 + b, x_0 + 2b, x_0 + 3b, \ldots.$$

That is,

$$x_n = x_0 + (n-1)b.$$

Hence, in (8), we shall assume that $a \neq 1$. From

$$x_{n+1} = ax_n + b \tag{9}$$

we get the associated homogeneous equation

$$x_{n+1} = ax_n. \tag{10}$$

Suppose $x_0 = A$ (a constant) in (10). Then

$$x_1 = aA$$
$$x_2 = a^2 A$$
$$x_3 = a^3 A$$

and, in general,

$$x_n = a^n A. \tag{11}$$

We now suppose that in (9), $x_n = B$ for every n. Substituting into (9), we have

$$B = aB + b$$
$$B = \frac{f}{b}1 - a \quad \text{(note that } a \neq 1).$$

That is,

$$x_n = \frac{b}{1-a} \tag{12}$$

is a particular solution of (9). A general solution of (9) is then given by adding the results in (11) and (12), or

$$x_n = a^n A + \frac{b}{1-a}. \tag{13}$$

Since we assume that x_0 is given, substitution of $n = 0$ into (13) yields

$$x_0 = A + \frac{b}{1-a}$$
$$A = x_0 - \frac{b}{1-a}.$$

Hence, (13) becomes

$$x_n = a^n \left[x_0 - \frac{b}{1-a} \right] + \frac{b}{1-a}. \tag{14}$$

We shall now show that the result in (14) does indeed offer a solution of (9) with x_0 given. We must show that (14) satisfies

$$x_{n+1} = ax_n + b. \tag{15}$$

Now, from (14) we have

$$x_{n+1} = a^{n+1} \left| x_0 - \frac{b}{1-a} \right| + \frac{b}{1-a}$$
$$= a \left(a^n \left[x_0 - \frac{b}{1-a} \right] \right) + \frac{b}{1-a},$$

while

$$ax_n + b = a \left(a^n \left[x_0 - \frac{b}{1-a} \right] + \frac{b}{1-a} \right) + b$$
$$= a \left(a^n \left[x_0 - \frac{b}{1-a} \right] \right) + \frac{ab}{1-a} + b$$
$$= a \left(a^n \left[x_0 - \frac{b}{1-a} \right] \right) + \frac{b}{1-a}.$$

Hence, the expression given in (14) does have the property that $x_{n+1} = ax_n + b$.

For pedagogical purposes, the student should learn the technique involved and should not try to memorize the formula given in (14).

Example: Consider the difference equation

$$x_{n+1} = .2x_n + .7 \tag{16}$$

with $x_0 = 3$. The associated homogeneous equation is

$$x_{n+1} = .2x_n.$$

For this equation, if $x_0 = A$ (a constant), then

$$x_1 = .2A$$
$$x_2 = (.2)^2 A$$
$$x_3 = (.2)^3 A$$

141

and, in general,

$$x_n = (.2)^n A. \tag{17}$$

To find a particular solution, we assume that in (16) $x_n = B$ for every n. Substituting, we have

$$B = .2B + .7$$
$$B = 7/8.$$

That is,

$$x_n = 7/8 \tag{18}$$

is a particular solution of (16).

Adding the results in (17) and (18), we obtain a general solution of (16) as

$$x_n = (.2)^n A + \frac{7}{8}. \tag{19}$$

We know that $x_0 = 3$. Substituting $n = 0$ into (19), we have

$$3 = A + \frac{7}{8}$$
$$A = \frac{17}{8}.$$

Hence, a solution of (16) with $x_0 = 3$ is given by

$$x_n = (.2)^n \left[\frac{17}{8}\right] + \frac{7}{8}.$$

Practice Problem: Use the method described in this appendix to find a solution to each of the following difference equations:

i) $x_{n+1} = 4x_n - 1, \ x_0 = 3$
ii) $x_{n+1} = .3x_n + .4, \ x_0 = .6$
iii) $x_{n+1} = x_n + .1, \ x_0 = 2$

Some After Dinner Reading Suggestions

For an extensive bibliography on applications of secondary mathematics, see one of the following:

Applications In School Mathematics, 1979 Yearbook, published by the National Council of Teachers of Mathematics, 1906 Association Drive, Reston, VA 22091

A Sourcebook of Applications of School Mathematics, published by a joint MAA/NCTM committee, available from the National Council of Teachers of Mathematics, 1906 Association Drive, Reston, VA 22091

For contemporary ideas in applied mathematics, read the following modules, developed and tested with support from the National Science Foundation, and produced by the Committee on Enrichment Modules, Dept. of Mathematics, University of Delaware, Newark, DE 19716:

APPLICATIONS OF MATHEMATICS TO MEDICINE (I) — Construction of Meaningful Indices, Addison-Wesley, Reading, Mass. (to appear).

APPLICATIONS OF MATHEMATICS TO MEDICINE (II) — Evaluation of Trauma Care, Janson Publications, Inc., Providence, R.I. (to appear).

CLUSTER ANALYSIS — with Applications, Janson Publications, Inc., Providence, R.I. (to appear).

DYNAMIC PROGRAMMING — An Approach to Solving Sequential Decision Problems, Addison-Wesley, Reading, Mass. (to appear).

GLYPHS — Graphical Representations of Multivariate Data, Addison-Wesley, Reading, Mass. (to appear).

GRAPH THEORY — with Applications, Addison-Wesley, Reading, Mass. (to appear).

GRAPHICAL ESTIMATION — Modern Developments in Curve Fitting, Janson Publications, Inc., Providence, R.I. (to appear).

INFORMATION THEORY — Application to Efficient Storage and Transmission of Information, Janson Publications, Inc., Providence, R.I. (to appear).

MATHEMATICAL THEORY OF SEARCH, Janson Publications, Inc., Providence, R.I. (to appear).

PATTERN RECOGNITION — with Applications, Janson Publications, Inc., Providence, R.I. (to appear).

QUEUES — An Analysis of Waiting Lines, Addison-Wesley, Reading, Mass. (to appear).

STATISTICAL BOOTSTRAPPING — The How and Why, Janson Publications, Inc., Providence, R.I. (to appear).